GW00372913

NT

Dictionary of
Physics

Dictionary of
Physics

NTC *Publishing Group*
Lincolnwood, Illinois USA

Library of Congress Cataloging-in-Publication Data

Dictionary of physics.
     p.   cm. -- (NTC pocket references)
    Originally published: Oxford England : Helicon Pub. Ltd., 1993.
    ISBN 0-8442-0923-6 (alk. paper)
    1. Physics--Dictionaries. I. Series
    QC5.D493 1996
    530'.03--dc20
                                       95-53968
                                            CIP

© 1996 by NTC Publishing Group, 4255 West Touhy Avenue,
Lincolnwood (Chicago), Illinois 60646-1975 U.S.A.
All rights reserved. Except for use in a review, the reproduction or utilization of the
work in any form or by any electronic, mechanical, or other means, now known or
hereafter invented, including scanning, photocopying, and recording, and in any
information storage and retrieval system, is forbidden without the written permis-
sion of the publisher. Original copyright © 1993 Helicon Publishing Ltd.,
Oxford, England.
Manufactured in the United Kingdom.

Editorial director
Michael Upshall

Consultant editor
John Avison BSc, CPhys, MInstP

Project editor
Sara Jenkins-Jones

Text editor
Catherine Thompson

Art editor
Terence Caven

Additional page make-up
Helen Bird

Production
Tony Ballsdon

absolute zero the lowest temperature theoretically possible, zero kelvin, equivalent to –273.16°C, at which molecules are motionless. Near absolute zero, the physical properties of some materials change substantially; for example, some metals lose their electrical resistance and become superconductive.

absorption the taking up of one substance by another, such as a liquid by a solid (ink by blotting-paper) or a gas by a liquid (ammonia by water). The term also refers to the phenomenon by which a substance retains radiation of particular wavelengths – for example, a piece of blue glass absorbs all visible light except the wavelengths in the blue part of the spectrum; it also describes the partial loss of energy resulting from the passage of electromagnetic and sound waves through a medium. In nuclear physics, absorption is the capture by elements such as boron of neutrons produced by fission in a reactor.

AC abbreviation for ◊*alternating current*.

acceleration the rate of change of the velocity of a moving body. It is measured in metres per second per second (m s^{-2}). Because velocity is a ◊vector quantity (possessing both magnitude and direction), a body travelling at constant speed may be said to be accelerating if its direction of motion changes. According to Newton's second law of motion, a body will only accelerate if it is acted upon by an unbalanced or resultant ◊force.

The average acceleration a of an object travelling in a straight line may be calculated using the formula:

$$a = \Delta v / \Delta t$$

where Δv is the change in velocity and Δt is the time taken, or by:

$$a = (v - u)/t$$

where u is the initial velocity of the object, v is its final velocity, and t is the time taken. A negative answer shows that the object is slowing down (decelerating). See also ◊equations of motion.

Acceleration due to gravity is the acceleration of a body falling freely under the influence of the Earth's gravitational field; it varies slightly at different latitudes and altitudes. The value adopted internationally for gravitational acceleration is 9.806 m s^{-2}.

acceleration, uniform acceleration in which the velocity of a body changes by equal amounts in successive time intervals. See ◊speed–time graph.

accommodation the ability of the eye to see objects clearly whether they are near or far away. The shape of the eye's lens is changed so that sharp images are focused on the ◊retina, becoming fatter (more biconvex) when focusing the image of a near object. From about the age of 40, the lens in the human eye becomes less flexible causing the defect of vision known as *presbyopia* or lack of accommodation. People with this defect need different spectacles for reading and distance vision.

accumulator storage ◊battery – that is, a group of rechargeable secondary cells.

An ordinary 12-volt car battery is an accumulator consisting of six lead–acid cells which are continually recharged by the car's alternator or dynamo. It has electrodes of lead and lead oxide in an electrolyte of sulphuric acid.

acoustics in general, the experimental and theoretical science of sound and its transmission; in particular, that branch of the science that has to do with the phenomena of sound in a particular space such as a room or theatre.

Acoustical engineering is concerned with the technical control of sound, and involves architecture and construction, studying control of ◊reverberation, soundproofing, and the elimination of noise; it also includes all forms of sound recording and reinforcement, the hearing and perception of sounds, and hearing aids.

action and reaction in mechanics, equal and opposite effects produced by a force acting on an object. For example, the pressure of expanding gases from the burning of fuel in a rocket engine (a force) produces an equal and opposite reaction, which causes the rocket to move. See ◊Newton's laws of motion.

activity the number of particles emitted in one second by a radioactive source. The term is used to describe the radioactivity or the potential danger of that source. The unit of activity is the becquerel (Bq).

ADC in electronics, abbreviation for ◊analogue-to-digital converter.

adder in electronics, the component of a digital system that carries out the process of adding two binary numbers. A separate adder is needed for each pair of binary ◊bits to be added. Each adder produces a sum and a carry bit.

advanced gas-cooled reactor (AGR) type of ◊nuclear reactor widely used in W Europe, especially Britain. The AGR uses a fuel of enriched uranium dioxide in stainless-steel cladding and a moderator of graphite. Carbon dioxide gas is pumped through the reactor core to extract the heat produced by the fission (see ◊nuclear fission) of the uranium. The heat is transferred to water in a steam generator, and the steam drives a turbogenerator to produce electricity.

In the UK, the first two AGR stations became operational at Hinkley Point and Hunterston in 1976. They each have a power output of 1320 megawatts from twin reactors.

aerial or *antenna* in radio and television broadcasting, a conducting device that radiates or receives electromagnetic waves. The design of an aerial depends principally on the wavelength of the signal. Long waves (hundreds of metres in wavelength) may require long wire aerials; short waves (several centimetres in wavelength) may require rods and dipoles; microwaves also may need dipoles—often with reflectors arranged like a toast rack—or highly directional parabolic dish aerials. Because microwaves travel in straight lines, giving line-of-sight communication, microwave aerials are usually located at the tops of tall masts or towers.

AGR abbreviation for ◊advanced gas-cooled reactor.

alloy metal blended with some other metallic or non-metallic substance to give it special qualities, such as resistance to corrosion, greater hardness or tensile strength, or stronger magnetism. Useful alloys include bronze, brass, cupronickel, solder, steel, and stainless steel.

alpha particle positively charged, high-energy particle emitted from the nucleus of a radioactive ◊atom. It is one of the products of the spontaneous disintegration of radioactive elements such as radium and thorium, and is identical with the nucleus of a helium atom—that is, it consists of two protons and two neutrons. The process of emission, *alpha decay*, transforms one element into another, decreasing the proton number by two and the nucleon number by four. See ◊radioactivity.

Because of their large mass, alpha particles have a short range of only a few centimetres in air, and can be stopped by a sheet of paper; they have a strongly ionizing effect (see ◊ionizing radiation) on the molecules that they strike, and are therefore capable of damaging living cells. Alpha particles travelling in a vacuum are deflected slightly by magnetic and electric fields.

alternating current (AC) electric current that flows for an interval of time in one direction and then in the opposite direction for another equal interval; that is, a current that flows in alternately reversed directions through or around a circuit. Electric energy is usually generated as alternating current in a power station, and alternating currents may be used for both power and lighting.

The advantage of alternating current over direct current (DC), as from a battery, is that its voltage can be raised or lowered economically by a transformer: high voltage for generation and transmission, and low voltage for safe use. Railways, factories, and domestic appliances, for example, use alternating current.

alternative energy energy from sources that are renewable and ecologically safe, as opposed to sources that are nonrenewable with toxic by-products, such as coal, oil, or gas (fossil fuels), and uranium (nuclear fuel). The most important alternative energy source is flowing water, harnessed as ◊hydroelectric power. Other sources include the ocean's tides and waves (see ◊tidal energy and ◊wave energy), the wind (◊wind energy), and the Sun (◊solar energy).

alternator an electricity ◊generator that produces an alternating current.

altimeter instrument used in aircraft that measures altitude, or height above sea level. The common type is a form of aneroid ◊barometer, which works by sensing the differences in air pressure at different altitudes. This must continually be recalibrated because of the change in air pressure with changing weather conditions. The ◊radar altimeter measures the height of the aircraft above the ground, measuring the time it takes for radio pulses emitted by the aircraft to be reflected. Radar altimeters are essential features of automatic and blind-landing systems.

AM abbreviation for ◊amplitude modulation.

ammeter instrument that measures electric current, usually in ◊amperes. The ammeter is placed in series with the component through which current is to be measured, and is constructed with a low internal resistance in order to prevent the reduction of that current as it flows through the instrument itself. A common type is the ◊moving- coil meter, which measures direct current (DC), but can, in the presence of a rectifier, measure alternating current (AC) also.

amp abbreviation for ◊*ampere*, a unit of electrical current.

ampere SI unit (abbreviation amp, symbol A) of electrical current. Electrical current is measured in a similar way to water current, in terms of an amount per unit time; one ampere represents a flow of about 6.28×10^{18} ◊electrons per second, or a rate of flow of charge of one coulomb per second.

Ampère's rule rule developed by André Ampère connecting the direction of an electric current and its associated magnetic field. It states that if a person were travelling along a current-carrying wire in the direction of conventional current flow (from the positive to the negative terminal), and carrying a magnetic compass, then the north pole of the compass needle would be deflected to the left-hand side.

amplifier electronic device that magnifies the strength of a signal, such as a radio signal. The ratio of the amplitude of the output signal to that of the input signal is called the ◊gain of the amplifier. As well as achieving high gain, an amplifier should be free from distortion and able to operate over a range of frequencies. Practical amplifiers are usually complex circuits, although simple amplifiers can be built from single transistors or valves. See also ◊voltage amplifier and ◊operational amplifier (op-amp).

amplitude maximum displacement of an oscillation from its equilibrium position. For a transverse wave motion, it is the height of a crest (or the depth of a trough); for a longitudinal wave, such as a sound wave, it is equal to half the distance between adjacent compressions (or adjacent rarefractions). The amplitude of a sound wave corresponds to the intensity (loudness) of the sound.

amplitude modulation (AM) one method by which radio waves are altered for the transmission of broadcasting signals. AM is constant in frequency, and varies the amplitude of the transmitting wave in accordance with the signal being broadcast.

analogue signal in electronics, a current or voltage that conveys or stores information, and varies continuously in the same way as the information it represents (compare ◊digital signal). Analogue signals are prone to interference and distortion.

The bumps in the grooves of an LP (gramophone) record form a mechanical analogue of the sound information stored, which is then is converted into an electrical analogue signal by the record player's pick-up device.

analogue-to-digital converter (ADC) in electronics, a circuit that converts an analogue signal into a digital signal. Such a circuit is used to convert the analogue electrical signal from a microphone into the digital signal that is stored on a compact disc.

AND gate in electronics, a type of ◊logic gate.

angle of declination angle at a particular point on the Earth's surface between the direction of the true or geographic North Pole and the magnetic north pole. The angle of declination has varied over time because of the slow drift in the position of the magnetic north pole.

analogue-to-digital converter

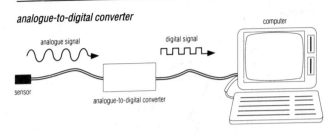

angle of dip or *angle of inclination* angle at a particular point on the Earth's surface between the direction of the Earth's magnetic field and the horizontal. It is measured using a *dip circle*, which has a magnetized needle suspended so that it can turn freely in the vertical plane of the magnetic field. In the northern hemisphere the needle dips below the horizontal, pointing along the line of the magnetic field towards the north pole. At the magnetic north and south poles, the needle dips vertically and the angle of dip is 90°.

angle of incidence angle between a ray of light striking a mirror (◊incident ray) and the normal to that mirror. It is equal to the ◊angle of reflection.

angle of reflection angle between a ray of light reflected from a mirror and the normal to that mirror. It is equal to the ◊angle of incidence.

angle of refraction angle between a refracted ray of light and the normal to the surface at which ◊refraction occurred. When a ray passes from air into a denser medium such as glass, it is bent towards the normal so that the angle of refraction is less than the ◊angle of incidence.

anion ion carrying a negative charge. During electrolysis, anions in the electrolyte move towards the anode (positive electrode).

annular eclipse solar ◊eclipse in which the Moon does not completely obscure the Sun and a thin ring of sunlight remains visible. Annular eclipses occur when the Moon is at its furthest point from the Earth.

angle of dip

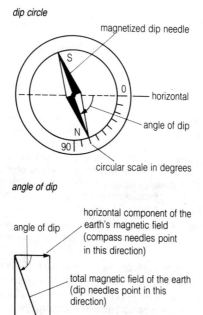

dip circle

magnetized dip needle

S

0 — horizontal

angle of dip

N

90

circular scale in degrees

angle of dip

horizontal component of the earth's magnetic field (compass needles point in this direction)

angle of dip

total magnetic field of the earth (dip needles point in this direction)

vertical component of the earth's magnetic field

anode the positive electrode towards which negative particles (anions, electrons) move within a device such as the cells of a battery, electrolytic cells, and diodes.

anomalous expansion of water the expansion of water as it is cooled from 4°C to 0°C. This behaviour is unusual, because most sub-

angle of incidence

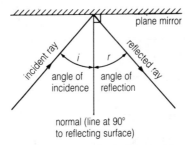

plane mirror

incident ray   *i*   *r*   reflected ray

angle of   angle of
incidence   reflection

normal (line at 90°
to reflecting surface)

stances contract when they are cooled. It means that water has a greater density at 4°C than at 0°C. Hence ice floats on water, and the water at the bottom of a pool in winter is warmer than at the surface. As a result large lakes freeze slowly in winter and aquatic life is more likely to survive.

antinode position in a ◊standing wave pattern at which the amplitude of vibration is greatest (compare ◊node). The standing wave of a stretched string vibrating in the fundamental mode has one antinode at its midpoint. A vibrating air column in a pipe has an antinode at the pipe's open end and at the place where the vibration is produced.

antinode

*the antinode on a vibrating string
for its fundamental mode*

antinode

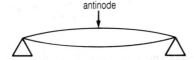

apparent depth depth that a transparent material such as water or glass appears to have when viewed from above. This is less than its real depth because of the ◊refraction that takes place when light passes into

apparent depth

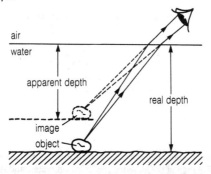

a less dense medium. The ratio of the real depth to the apparent depth of a transparent material is equal to its ◊refractive index.

Archimedes' principle law stating that an object that is totally or partly submerged in a fluid displaces a volume of fluid that weighs the same as the apparent loss in weight of the object (which equals the upthrust on it).

 If the weight of the object is less than the force exerted by the fluid, it will float partly or completely above the surface; if its weight is equal to the force, the object will come to equilibrium below the surface. See ◊floating.

armature in a motor or generator, the wire-wound coil that carries the current and rotates in a magnetic field. (In alternating-current machines, the armature is sometimes stationary.) The term also refers to the pole piece of a permanent magnet or electromagnet. The moving, iron part of a solenoid, especially if it acts as a switch, may also be called an armature.

artificial radioactivity radioactivity arising from synthetic ◊radioisotopes (radioactive isotopes or elements that are formed when elements are bombarded with subatomic particles—protons, neutrons, or electrons—or small nuclei).

atmosphere the mixture of gases that surrounds the Earth; it is prevented from escaping by the pull of the Earth's gravity. ◊Atmospheric pressure decreases with height in the atmosphere.

The lowest level of the atmosphere, the *troposphere*, is heated by the Earth, which is warmed in turn by infrared and visible radiation from the Sun. Warm air cools as it rises in the troposphere, causing rain and most other weather phenomena. The upper levels of the atmosphere, particularly the *ozone layer*, absorb almost all of the ultraviolet light radiated by the Sun, and prevent lethal amounts from reaching the Earth's surface.

atmospheric pressure the ◊pressure at any point in the atmosphere that is due to the weight of air above it; it therefore decreases with height. At sea level the pressure is about 101 kilopascals; however, the exact value varies according to temperature and weather. Changes in atmospheric pressure, measured with a ◊barometer, are used in weather forecasting.

atom the smallest unit of matter that can take part in a chemical reaction, and which cannot be broken down chemically into anything simpler. An atom is made up of protons and neutrons in a central nucleus surrounded by electrons (see ◊atomic structure). The atoms of the various elements differ in proton number, relative atomic mass, and behaviour. There are 109 different types of atom, corresponding with the 109 known elements.

Atoms are much too small to be seen even by the microscope (the largest, caesium, has a diameter of 0.0000005 mm), and they are in constant motion. Belief in the existence of atoms dates back to the ancient Greek natural philosophers. The first scientist to gather evidence for the existence of atoms was John Dalton, in the 19th century, who believed that every atom was a complete unbreakable entity. In the early 20th century, Ernest Rutherford showed by experiment that an atom in fact consists of a nucleus surrounded by negatively charged particles called electrons.

atomic energy another name for ◊nuclear energy.

atomic mass see ◊relative atomic mass.

atomic mass unit or *dalton* unit (symbol amu or u) of mass that is used to measure the relative mass of atoms and molecules. It is equal to one-twelfth of the mass of a carbon-12 atom, which is equivalent to the mass of a proton or 1.66×10^{-27} kg. The ◊relative atomic mass of an atom has no units; thus oxygen-16 has an atomic mass of 16 daltons, but a relative atomic mass of 16.

atomic number alternative name for the ◊proton number of an atom.

atomic physics the study of the structure and the properties of the ◊atom.

atomic radiation or *nuclear radiation* energy given out by disinte-grating atoms during ◊radioactive decay. The energy may be in the form of fast-moving particles, known as ◊alpha particles and ◊beta par-ticles, or in the form of high-energy electromagnetic waves known as ◊gamma radiation. Overlong exposure to atomic radiation can lead to radiation sickness. Radiation biology studies the effect of radiation on living organisms.

atomic structure the internal structure of an ◊atom. The core of the atom is the *nucleus*, a particle only one ten-thousandth the diameter of the atom itself. The simplest nucleus, that of hydrogen, comprises a single positively charged particle, the *proton*. Nuclei of other elements contain more protons and additional particles of about the same mass as the proton but with no electrical charge, *neutrons*. Each element has its own characteristic nucleus with a unique number of protons, the proton (or atomic) number. The number of neutrons may vary. Where atoms of a single element have different numbers of neutrons, they are called ◊isotopes. Although some isotopes tend to be unstable and exhibit ◊radioactivity, they all have identical chemical properties.

The nucleus is surrounded by a number of *electrons*, each of which has a negative charge equal to the positive charge on a proton, but which weighs only 1/1839 times as much. For a neutral atom, the nucleus is surrounded by the same number of electrons as it contains protons. The chemical properties of an element are determined by the ease with which its atoms can gain or lose electrons.

Atoms are held together by the electrical forces of attraction between each negative electron and the positive protons within the nucleus. The latter repel one another with relatively enormous forces; a nucleus holds together only because other forces, not of a simple electrical character, attract the protons and neutrons to one another. These additional forces act only so long as the protons and neutrons are virtually in contact with one another. If, therefore, a fragment of a complex nucleus, containing some protons, becomes only slightly loosened from the main group of neutrons and protons, the strong natural repulsion between the protons will cause this fragment to fly apart from the rest of the nucleus at high speed. It is by such fragmentation of atomic nuclei (◊nuclear fission) that nuclear energy is released.

B

background radiation radiation that is always present in the environment. By far the greater proportion (87%) of it is emitted from natural sources. Alpha and beta particles and gamma radiation are radiated by the traces of radioactive minerals that occur naturally in the ground, the underlying rocks, the bricks and stones of buildings, and even in the human body, and by radioactive gases such as radon and thoron, which are found in soil and may seep upwards into buildings. Radiation from space (cosmic radiation) also contributes to the background level. Because background radiation increases the count rate obtained during any experimental work on radioactive materials, a background count rate is usually subtracted from any data obtained.

balance state of equilibrium achieved by an object when the forces acting on it (or the moments of those forces) cancel each other. The resultant force acting on a balanced object is zero.

balance apparatus for weighing or measuring mass. The various types include the ◊beam balance, consisting of a centrally pivoted lever with pans hanging from each end, and the ◊spring balance, in which the object to be weighed stretches (or compresses) a vertical coil spring. Kitchen and bathroom scales are balances.

barometer instrument that measures ◊atmospheric pressure as an indication of weather. Most often used are the *mercury barometer* and the *aneroid barometer*.

In a mercury barometer a column of mercury in a glass tube roughly 0.75 m high (closed at one end, curved upwards at the other) is balanced by the pressure of the atmosphere on the open end; any change in the height of the column reflects a change in pressure. An aneroid barometer achieves a similar result by changes in the distance between the faces of a shallow cylindrical metal box, from which most of the air has been removed.

barometer

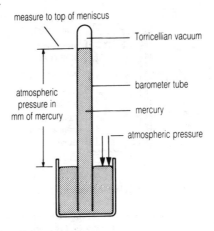

measure to top of meniscus

Torricellian vacuum

barometer tube

atmospheric
pressure in
mm of mercury

mercury

atmospheric pressure

aneroid barometer

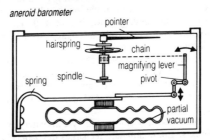

pointer

hairspring

chain

magnifying lever

spring   spindle

pivot

partial
vacuum

battery any energy storage device allowing release of electricity on
demand. A battery is made up of one or more *cells*, each containing two
conducting ()electrodes, one positive and one negative, immersed in an
electrolyte (an electrically conducting solution or molten substance)
in a container. When an outside connection, such as through a light
bulb, is made between the electrodes, a current flows through the cir-

battery

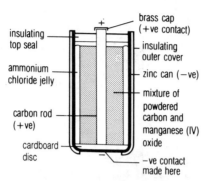

cuit, and chemical reactions releasing energy take place within the cells.

Primary-cell batteries are disposable; *secondary-cell*, or storage, batteries are rechargeable. The common *dry cell* is a primary-cell battery consisting of a central carbon electrode immersed in a paste of manganese dioxide and ammonium chloride that acts as the electrolyte. The zinc casing forms the other electrode. It is dangerous to try to recharge a primary-cell battery.

The introduction of rechargeable nickel-cadmium batteries has revolutionized portable electronic newsgathering (sound recording, video) and information processing (computing). These batteries offer a stable, short-term source of power free of noise and other hazards associated with mains electricity.

The lead–acid *car battery* is a secondary-cell battery, or accumulator. The car's generator continually recharges the battery, which consists of sets of lead (positive) and lead peroxide (negative) plates in an electrolyte of sulphuric acid.

beam balance instrument for measuring mass (or weight). A simple form consists of a beam pivoted at its midpoint with a pan hanging at each end. The mass to be measured, in one pan, is compared with a variety of standard masses placed in the other. When the beam is balanced, the masses' turning effects or moments under gravity, and hence the masses themselves, are equal.

beat regular variation in the loudness of a sound when two notes of nearly equal pitch or ◊frequency are heard together. Beats result from the ◊interference between the sound waves of the notes. The frequency of the beats equals the difference in frequency of the notes.

Musicians use the effect when tuning their instruments. A similar effect can occur in electrical circuits when two alternating currents are present, producing regular variations in the overall current.

becquerel SI unit (symbol Bq) of ◊radioactivity, equal to one radioactive disintegration (change in the nucleus of an atom when a particle or ray is given off) per second.

Bernoulli's principle statement that the speed of a fluid varies inversely with the pressure; an increase in speed produces a decrease in pressure (such as the drop in hydraulic pressure experienced when a fluid flows more rapidly through a constriction in a pipe) and vice versa. The principle also explains the pressure differences on each surface of an aerofoil, which gives lift to the wing of an aircraft. The principle was named after the Swiss physicist Daniel Bernoulli.

beta particle electron ejected with great velocity from a radioactive atom that is undergoing spontaneous disintegration. Beta particles do not exist in the nucleus but are created on disintegration, *beta decay*, when a neutron converts to a proton to emit an electron. The process transforms one element into another, increasing the proton number by one while the nucleon number remains the same.

Beta particles are more penetrating than ◊alpha particles, but less so than ◊gamma radiation; they can travel several metres in air, but are stopped by 2–3 mm of aluminium. They are less strongly ionizing than alpha particles and, like cathode rays, are easily deflected by magnetic and electric fields.

bimetallic strip strip made from two metals each having a different coefficient of ◊thermal expansion; it therefore bends when subjected to a change in temperature. Such strips are used widely for temperature measurement and control.

binoculars magnifying optical instrument for viewing an object with both eyes. Binoculars consist of two telescopes containing convex lenses and prisms, which together produce a stereoscopic, seemingly three-dimensional effect as well as magnifying the image. The use of prisms has the effect of 'folding' the light path, allowing for a compact design.

binocular vision vision in which both eyes can focus on an object at the same time. In humans, the eyes, which are about 7 cm apart, provide two slightly different images of the world, but these are coordinated to give a three-dimensional perception that allows the brain to judge accurately the position and speed of objects up to 60 m away.

bistable circuit or *flip-flop* in electronics, a simple circuit composed of two NAND ◊logic gates that remains in one of two stable states until it receives a pulse (logic 1 signal) through one of its inputs, upon which it switches, or 'flips', over to the other state. It is used as a storage device, and is usually found incorporated in an ◊integrated circuit.

bit in computing, the smallest unit of information; a binary digit or place in a binary number. A ◊byte is eight bits.

block and tackle type of ◊pulley.

Bohr model model of the atom conceived by Neils Bohr 1913. It assumes that the following rules govern the behaviour of electrons: (1) electrons revolve in orbits of specific radius around the nucleus without emitting radiation; (2) within each orbit, each electron has a fixed amount of energy; electrons in orbits farther away from the nucleus have greater energies; (3) an electron may 'jump' from one orbit of high energy to another of lower energy causing the energy difference to be emitted as a ◊photon of electromagnetic radiation such as light. The Bohr model has been superseded by wave mechanics (see ◊quantum theory).

boiling rapid conversion of a liquid into vapour that takes place when the liquid reaches a certain temperature (◊boiling point). It involves the formation of vapour bubbles within the body of a liquid, whereas ◊evaporation occurs only at the surface.

boiling point for any given liquid, the temperature at which the application of heat raises the temperature of the liquid no further, but converts it to vapour.

The boiling point of water under normal pressure is 100°C. The lower the pressure, the lower the boiling point and vice versa.

Bourdon gauge instrument for measuring ◊pressure, invented by Eugene Bourdon 1849. The gauge contains a C-shaped tube, closed at one end. When the pressure inside the tube increases, the tube uncurls slightly, causing a small movement at its closed end. A system of levers and gears magnifies this movement and turns a pointer, which indicates the pressure on a circular scale. Bourdon gauges are often fitted to cylinders of compressed gas used in industry and hospitals.

Bourdon gauge

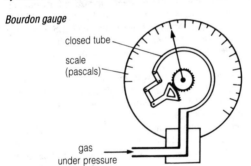

closed tube

scale
(pascals)

gas
under pressure

Boyle's law law stating that the volume of a given mass of gas at a constant temperature is inversely proportional to its pressure. For example, if the pressure of a gas doubles, its volume will be reduced by a half, and vice versa. The law was discovered in 1662 by Robert Boyle.

Boyle's law may be expressed as:

$$\text{pressure} \times \text{volume} = \text{constant}$$

or, more usefully, as:

$$P_1 V_1 = P_2 V_2$$

where P_1 and V_1 are the initial pressure and volume of a gas, and P_2 and V_2 are its final pressure and volume. See also ◊gas laws.

braking distance distance travelled by a vehicle while the brakes are being applied at maximum efficiency. The ◊stopping distance travelled by a vehicle is the sum of the braking distance and the ◊thinking distance.

breeder reactor or *fast breeder* alternative names for ◊fast reactor, a type of nuclear reactor.

bridge structure that provides a continuous path or road over water, valleys, ravines, or above other roads. Bridges can be designed according to four principles: *arch* for example, Sydney Harbour bridge (steel arch), Australia, with a span of 503 m; *beam or girder* for example, Rio–Niteroi, Guanabara Bay, Brazil, centre span 300 m; length 13,900 m; *cantilever* for example, Forth rail bridge, Scotland, 1,658 m long with two main spans, two cantilevers each, one from each tower; *suspension* for example, Humber bridge, England, with a centre span of 1,410 m.

The types of bridge differ in the way they bear the weight of the structure and its load. Beam, or girder, bridges are supported at each end by the ground with the weight thrusting downwards. Cantilever bridges are a complex form of girder. Arch bridges thrust outwards but downwards at their ends; they are in compression. Suspension bridges use cables under tension to pull inwards against anchorages on either side of the span, so that the roadway hangs from the main cables by the network of vertical cables. Some bridges are too low to allow traffic to pass beneath easily, so they are designed with moveable parts, like swing and draw bridges.

brittle material material that breaks suddenly under stress at a point just beyond its elastic limit (see ◊elasticity). Brittle materials may also break suddenly when given a sharp knock. Pottery, glass, and cast iron are examples of brittle materials. Compare ◊ductile material.

Brownian movement the continuous random motion of particles in a fluid medium (gas or liquid) as they are subjected to impact from the molecules of the medium. The phenomenon was explained by Albert

Einstein in 1905 but was observed as long ago as 1827 by the Scottish botanist Robert Brown.

brush in certain electric motors, one of a pair of contacts that pass electric current into and out of the rotating coils by means of a device known as a ◊commutator. The brushes, which are often replaceable, are usually made of a soft, carbon material to reduce wear of the copper contacts on the rotating commutator.

bubble chamber device for observing the nature and movement of atomic particles. It consists of a vessel filled with a superheated liquid through which ionizing particles move and collide. The paths of these particles are shown by strings of bubbles, which can be photographed and studied. It was invented by Donald Glaser in 1952. Compare ◊cloud chamber.

byte in computing, a basic unit of storage of information. A byte contains eight ◊bits and can hold either a single character (letter, digit, or punctuation symbol) or a number between 0 and 255.

C

capacitor or *condenser* device for storing electric charge, used in electronic circuits; it consists of two or more metal plates separated by an insulating layer called a dielectric.

Its *capacitance* is the ratio of the charge stored on either plate to the potential difference between the plates. The SI unit of capacitance is the farad, but most capacitors have much smaller capacitances, and the microfarad (a millionth of a farad) is the commonly used practical unit.

capillarity the spontaneous movement of liquids up or down narrow tubes, or capillaries. The movement is due to unbalanced molecular attraction at the boundary between the liquid and the tube. If liquid molecules near the boundary are more strongly attracted to molecules in the material of the tube than to other nearby liquid molecules, the liquid will rise in the tube. If liquid molecules are less attracted to the material of the tube than to other liquid molecules, the liquid will fall.

cathode the negative electrode towards which positive particles (cations) move within a device such as a cell in a battery or an electrolytic cell.

cathode-ray oscilloscope (CRO) instrument used to measure electrical potentials or voltages that vary over time and to display the waveforms of electrical oscillations or signals. Readings are displayed graphically on the screen of a cathode-ray tube.

cathode rays streams of fast-moving electrons that travel from a cathode (negative electrode) towards an anode (positive electrode) in a vacuum tube. They carry a negative charge and can be deflected by electric and magnetic fields. Cathode rays focused into fine beams of fast electrons are used in cathode-ray tubes, the electrons' ◊kinetic energy being converted into light energy as they collide with the tube's fluorescent screen. See also ◊Maltese-cross tube.

cation ◊ion carrying a positive charge. During electrolysis, cations in the electrolyte move to the cathode (negative electrode).

cell, electric apparatus in which chemical energy is converted into electrical energy; the popular name is 'battery', but this actually refers to a collection of cells in one unit. A *primary* electric cell cannot be replenished, whereas in a *secondary* cell or storage battery, the action is reversible and the original condition can be restored by an electric current. The first cell was made by Alessandro Volta in 1800. See also ◊battery.

Celsius temperature scale in which one division or degree is taken as one hundredth part of the interval between the freezing point (0°C) and the boiling point (100°C) of water at standard atmospheric pressure.

centre of curvature in optics, the centre of the sphere of which a spherical mirror is part.

centre of mass or *centre of gravity* the point in or near an object from which its total weight appears to originate and can be assumed to act. A symmetrical homogeneous object such as a sphere or cube has its centre of mass at its physical centre; a hollow shape (such as a cup) may have its centre of mass in space inside the hollow. See also ◊stability.

centripetal force force that acts radially inwards on an object moving in a curved path. For example, with a weight whirled in a circle at the end of a length of string, the centripetal force is the tension in the string. For an object of mass m moving with a velocity v in a circle of radius r, the centripetal force F is given by:

$$F = mv^2/r$$

The reaction to this force is the *centrifugal force*.

chain reaction in nuclear physics, a fission reaction that is maintained because neutrons released by the splitting of some atomic nuclei themselves go on to split others, releasing even more neutrons. Such a reaction can be controlled (as in a nuclear reactor) by using moderators to absorb excess neutrons. Uncontrolled, a chain reaction produces a nuclear explosion (as in an atom bomb).

change of state a change in the physical state (solid, liquid, or gas) of a material. See ◊state change.

charge see ◊electric charge.

Charles's law law stating that the volume of a given mass of gas at constant pressure is directly proportional to its absolute temperature (temperature in kelvin). It was discovered by Jacques Charles in 1787, and independently by Joseph Gay-Lussac in 1802.

Charles's law may be expressed as:

$$\text{volume/temperature} = \text{constant}$$

or, more usefully, as:

$$V_1/T_1 = V_2/T_2$$

where V_1 and T_1 are the initial volume and temperature in kelvin of a gas, and V_2 and T_2 are its final volume and temperature. See also ◊gas laws.

choke coil employed as an electrical inductor, particularly the type used as a 'starter' in the circuit of fluorescent lighting.

ciliary body ring of muscle inside the vertebrate eye that controls the shape of the lens, allowing images of objects at different distances to be focused on the retina. When the muscle is relaxed the lens has its longest ◊focal length and focuses rays from distant objects. Contraction of the muscle reduces tension in the lens, making it more curved; the lens therefore has a shorter focal length and focuses images of near objects.

circuit an arrangement of electrical components through which a current can flow. There are two basic types, ◊series circuits and ◊parallel circuits. In a series circuit, the components are connected end-to-end so that the current flows through all components one after the other. In a parallel circuit, components are connected side-by-side so that part of the current is divided through each component.

circuit breaker switching device designed to protect an electric circuit from excessive current. It has the same action as a ◊fuse, and many houses now have a circuit breaker between the incoming mains supply

circuit diagram

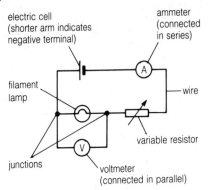

and the domestic circuits. Circuit breakers usually work by means of ◊solenoids. Those at electricity-generating stations have to be specially designed to prevent dangerous arcing (the release of luminous discharge) when the high-voltage supply is switched off.

circuit diagram simplified drawing of an electric circuit. The circuit's components are represented by internationally recognized symbols, and the connecting wires by straight lines. A dot indicates where wires join.

circular motion motion in a circular path or orbit of a particle or body. The resultant, or unbalanced, force responsible is the ◊centripetal force, and is directed towards the centre of the path.

cloud chamber apparatus for tracking ionized particles. It consists of a vessel fitted with a piston and filled with air or other gas, supersaturated with water vapour. When the volume of the vessel is suddenly expanded by moving the piston outwards, the vapour cools and a cloud of tiny droplets forms on any nuclei, dust, or ions present. As single fast-moving ionizing particles collide with the air or gas molecules, they show as visible tracks. See also ◊bubble chamber.

CMOS abbreviation for *complementary metal-oxide semiconductor*, a family of ◊integrated circuits widely used in building electronic systems because they can operate at any voltage between 3 and 15 volts, and have low power consumption and heat dissipation; however, CMOS circuits have lower operating speeds than have circuits of the ◊TTL, or transistor–transistor logic, family.

When building electronic systems it is usually advisable to build using integrated circuits all from one family.

colour quality or wavelength of light emitted or reflected from an object. Visible white light consists of electromagnetic radiation of various wavelengths, and if a beam is refracted through a prism, it can be spread out into a spectrum, in which the various colours correspond to different wavelengths. From long to short wavelengths (from about 700 to 400 nanometres) the colours are red, orange, yellow, green, blue, indigo, and violet.

When a surface is illuminated, some parts of the white light are absorbed, depending on the molecular structure of the material and the dyes applied to it. A surface that looks red absorbs light from the blue end of the spectrum, but reflects light from the red, long-wave end. Colours vary in brightness, hue, and saturation (the extent to which they are mixed with white).

colour code standard system of coding in which colours are used to indicate a value of an electrical component or to identify a particular wire in a cable.

The *resistor colour code* indicates the resistance in ohms of a resistor in an electrical circuit, usually by means of four bands of colour painted round the resistor. In Europe the standard *wire colour code* for wires connecting equipment to the mains supply is brown insulator for the live wire, blue for the neutral wire, and green with a yellow stripe for the earth wire.

colour vision the ability of the eye to recognize different frequencies in the visible spectrum as colours. In most vertebrates, including humans, colour vision is due to the presence on the ◊retina of three types of light-sensitive cone cell, each of which responds to a different primary colour (red, green, or blue).

commutator device in a DC (direct-current) electric motor that reverses the current flowing in the armature coils as the armature rotates. A DC generator, or ◊dynamo, uses a commutator to convert the AC (alternating current) generated in the armature coils into DC. A commutator consists of opposite pairs of conductors insulated from one another, and contact to an external circuit is provided by carbon or metal brushes.

compact disc disc for digital information storage, about 12 cm across, mainly used for music, when it has up to an hour's playing time on one side. The compact disc is made of aluminium with a transparent plastic coating; the metal disc underneath is etched by a ◊laser beam with microscopic pits that carry a digital code representing the sounds. During playback, a laser beam reads the code and produces signals that are changed into near-exact replicas of the original sounds.

compass any instrument for finding direction. The most commonly used is the ◊magnetic compass.

component in mechanics, one of two or more forces (or other ◊vector quantities) acting on a body whose combined effect is represented by a single ◊resultant force. For purposes of calculation, a force may be resolved (see ◊resolution of forces) into components acting at right angles to each other.

computer a programmable electronic device that processes data and performs calculations and other manipulation tasks. There are three types: the *digital computer*, which manipulates information coded as binary numbers, the *analogue computer*, which works with continuously varying quantities, and the *hybrid computer*, which has characteristics of both analogue and digital computers.

There are four sizes of digital computer, corresponding roughly to their memory capacity and processing speed. *Microcomputers* are the smallest and most common, used in small businesses, at home, and in schools. They are usually single-user machines. *Minicomputers* are found in medium-sized businesses and university departments. They may support from a dozen to 30 or so users at once. *Mainframes*, which can often service several hundreds of users simultaneously, are found in large organizations such as national companies and government departments. *Supercomputers* are mostly used for highly complex scientific tasks, such as analysing the results of nuclear physics experiments and weather forecasting.

basic components At the heart of a computer is the *CPU* (central processing unit), which performs all the computations. This is supported by memory, which holds the current program and data, and 'logic arrays', which help move information around the system. The computer's 'device driver' circuits control input and output devices, such as keyboards, VDU screens, printers, and disc-drive units for mass memory storage.

concave lens lens that possesses at least one surface that curves inwards. It is a ◊diverging lens, spreading out those light rays that have been refracted through it. A concave lens is thinner at its centre than at its edges.

Common forms include the *biconcave* lens (with both surfaces curved inwards) and the *plano-concave* (with one flat surface and one concave). The whole lens may be further curved overall, and this is called a *convexo-concave* or diverging meniscus lens, as in some lenses used for corrective purposes.

concave mirror curved or spherical mirror that reflects light from its inner surface. It may be either circular or parabolic in section. A concave mirror converges light rays to form a reduced, inverted, real image in front, or an enlarged, upright, virtual image seemingly behind it, depending on how close the object is to the mirror.

concave mirror

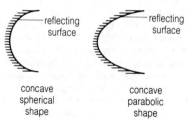

concave
spherical
shape

concave
parabolic
shape

Only a parabolic concave mirror has a true, single-point ◊principal focus for parallel rays. For this reason, parabolic mirrors are used as reflectors to focus light in telescopes, or to focus microwaves in satellite communication systems. The reflector behind a spot lamp or car headlamp is parabolic.

condensation the conversion of a vapour to a liquid as it loses heat. This is frequently achieved by letting the vapour come into contact with a cold surface.

condenser in optics, a short-focal-length convex ◊lens or combination of lenses used for concentrating a light source on to a small area, as used in a slide projector or microscope substage lighting unit. A condenser can also be made using a concave mirror. In electricity, it is another name for a ◊capacitor.

conduction, electrical the flow of charged particles through a material that gives rise to electric current. Conduction in metals involves the flow of negatively charged free ◊electrons. Conduction in gases and some liquids involves the flow of ◊ions that carry positive charges ions in one direction and negative charges in the other. Conduction in a semiconductor such as silicon involves the flow of electrons and positive holes.

conduction, heat flow of heat energy through a material without the movement of any part of the material itself (compare ◊conduction, electrical). Heat energy is present in all materials in the form of the kinetic energy of their vibrating molecules, and may be conducted from one molecule to the next in the form of this mechanical vibration. In the case of metals, which are particularly good conductors of heat, the free electrons within the material carry heat around very quickly.

conductivity, heat measure of how well a material conducts heat. A good conductor, such as a metal, has a high conductivity; a poor conductor, called an insulator, has a low conductivity. See also ◊U-value.

conductor material that conducts heat or electricity (as opposed to a non-conductor or insulator). A good conductor has a high electrical or heat conductivity, and is generally a substance rich in free electrons such as a metal. A poor conductor (such as the non-metals glass and porcelain) has few free electrons. Carbon is exceptional in being non-metallic and yet (in some of its forms) a relatively good conductor of heat and electricity. Substances such as silicon and germanium, with intermediate conductivities that are improved by heat, light, or voltage, are known as ♢semiconductors.

conservation of momentum in mechanics, a law that states that total ♢momentum is conserved (remains constant) in all collisions, providing no external resultant force acts on the colliding bodies. The principle may be expressed as an equation used to solve numerical problems:

total momentum before collision = total momentum after collision.

contact force force or push produced when two objects are pressed together and their surface atoms try to keep them apart. Contact forces always come in pairs—for example, the downwards force exerted on a floor by the sole of a person's foot is matched by an equal upwards force exerted by the floor on that sole.

convection current movement of heat energy through a liquid or gas that involves the flow of the medium itself (compare heat ♢conduction). Convection is caused by the expansion of the medium as its temperature rises; the expanded material, being less dense, rises above colder and therefore denser material. In some heating systems, convection currents are used to carry hot water upwards in pipes. Such currents arise in the atmosphere above hot land masses or warm seas causing breezes.

conventional current direction in which an electric current is considered to flow in a circuit. By convention, the direction is that in which positive-charge carriers would flow—from the positive terminal of a cell to its negative terminal. In circuit diagrams, the arrows shown on symbols for components such as diodes and transistors point in the direction of conventional current flow.

converging lens lens that converges or brings to a focus those light rays that have been refracted by it. It is a ♢convex lens, with at least one surface that curves outwards, and is thicker towards the centre than at the edge. Converging lenses are used to form real images in many

convection current

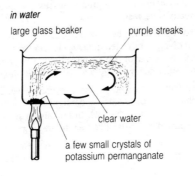

in water

large glass beaker   purple streaks

clear water

a few small crystals of
potassium permanganate

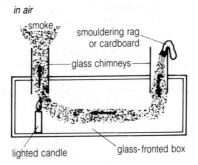

in air

smoke

smouldering rag
or cardboard

glass chimneys

lighted candle   glass-fronted box

◊optical instruments; those that form virtual, magnified images are
used as ◊magnifying glasses or to correct ◊long-sightedness.

convex lens lens that possesses at least one surface that curves outwards. It is a ◊converging lens, bringing rays of light to a focus. A convex lens is thicker at its centre than at its edges.

Common forms include the *biconvex* lens (with both surfaces
curved outwards) and the *plano-convex* (with one flat surface and one

convex). The whole lens may be further curved overall, in what is called a ***concavo-convex*** or converging meniscus lens, as in some lenses used in corrective eyewear.

convex mirror curved or spherical mirror that reflects light from its outer surface. It diverges reflected light rays to form a reduced, upright, virtual image. Convex mirrors give a wide field of view and are therefore particularly suitable for surveillance purposes in shops.

cornea transparent front section of the vertebrate ◊eye. The cornea is curved and behaves as a fixed lens, so that light entering the eye is partly focused before it reaches the lens. In humans, diseased or opaque parts may be replaced by grafts of corneal tissue from a donor.

coulomb SI unit (symbol C) of electrical charge. One coulomb is the quantity of electricity conveyed by a current of one ampere in one second.

count rate number of particles emitted per unit time by a radioactive source. It is measured by a counter, such as a ◊Geiger counter, or ◊ratemeter.

couple in mechanics, a pair of forces acting on an object that are equal in magnitude and opposite in direction, but do not act along the same straight line. The two forces produce a turning effect or ◊moment that tends to rotate the object; however, no single resultant (unbalanced) force is produced and so the object is not moved from one position to another.

The moment of a couple is the product of the magnitude of either of the two forces and the perpendicular distance between those forces. If

couple

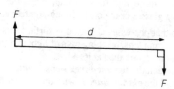

the magnitude of the force is F newtons and the distance is d metres then the moment, in newton-metres, is given by:

$$\text{moment} = Fd$$

critical angle in optics, for a ray of light passing from a denser to a less dense medium (such as from glass to air), the smallest angle of incidence at which the emergent ray grazes the surface of the denser medium—at an angle of refraction of 90°.

When the angle of incidence is less than the critical angle, the ray passes out (is refracted) into the less dense medium; when the angle of incidence is greater than the critical angle, the ray is reflected back into the denser medium.

critical angle

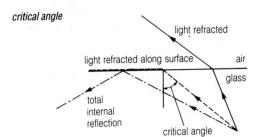

critical mass in nuclear physics, the minimum mass of fissile material that can undergo a continuous ◊chain reaction. Below this mass, too many ◊neutrons escape from the surface for a chain reaction to carry on; above the critical mass, the reaction may accelerate into a nuclear explosion.

critical reaction in a nuclear physics, a self-sustaining chain reaction in which the number of neutrons being released by the nuclear fission of uranium–235 nuclei and the number of neutrons being absorbed by uranium–238 nuclei and by control rods are balanced. If balance is not achieved the reaction will either slow down and cease to generate enough power, or will build up and go out of control, as in a nuclear explosion. Control rods are used to adjust the rate of reaction and maintain balance.

critical temperature temperature above which a particular gas cannot be converted into a liquid by pressure alone. It is also the temperature at which a magnetic material loses its magnetism (the Curie temperature or point).

CRO abbreviation for ◊*cathode-ray oscilloscope*.

CRT abbreviation for *cathode-ray tube* (see ◊cathode rays).

crumple zone region at the front and rear of a motor vehicle that is designed to crumple gradually during a collision, so reducing the risk of serious injury to passengers. The progessive crumpling absorbs the kinetic energy of the vehicle more gradually than would a rigid structure, thereby diminishing the forces of deceleration acting on the vehicle and on the people inside.

The crumple zone's effect is based on the principle that the ◊impulse required to stop a vehicle and reduce its momentum to zero is equal to the product of the decelerating force and the time over which that force acts. It follows that if the length of time is increased, the force will be reduced.

crystalline solid solid in which the atoms, ions, or molecules are arranged in a regularly repeating three-dimensional pattern giving it a permanent and rigid shape. The pattern or array extends continuously to form *crystals*, which have a characteristic geometric form determined by the packing arrangement of their molecules, and may vary in size from the sub-microscopic to structures some 30 m in length. Three common crystalline forms are: (1) the simple cubic structure of ionic crystals, such as those of sodium chloride (NaCl); (2) the face-centred cubic structure of metals such as aluminium, copper, gold, silver, and lead; and (3) the hexagonal close-packed structure of metals such as cadmium and zinc.

current, electric the flow of electrically charged particles through a conducting circuit due the presence of a ◊potential difference. The current at any point in a circuit is the amount of charge flowing per second; it is measured in amperes (coulombs per second). If the amount of charge is Q coulombs and the time over which charge flow is measured is t seconds then the current I is given by the formula:

$$I = Q/t$$

Current carries electrical energy from a power supply, such as a battery of electrical cells, to the components of the circuit where it is converted into other forms of energy, such as heat, light, or motion. It may be either direct (DC, see ◊direct current) or alternating (AC, see ◊alternating current). See also ◊Ohm's law.

heating effect When current flows in a component possessing resistance, electrical energy is converted into heat. If the resistance of the component is R ohms and the current through it is I amperes, then the heat energy W (in joules) generated in a time t seconds is given by the formula:

$$W = I^2Rt$$

magnetic effect A ◊magnetic field is created around all conductors that carry a current. When a current-bearing conductor is made into a coil it forms an ◊electromagnet with a magnetic field that is similar to that of a bar magnet, but which disappears as soon as the current is switched off. The strength of the magnetic field is directly proportional to the current in the conductor – a property that allows a small electromagnet to be used to produce a pattern of magnetism on recording tape that accurately represents the sound or data to be stored. The direction of the field created around a conducting wire may be predicted by using ◊Maxwell's screw rule.

motor effect A conductor carrying current in a magnetic field experiences a force, and is impelled to move in a direction perpendicular to both the direction of the current and the direction of the magnetic field. The direction of motion may be predicted by Fleming's left-hand rule (see ◊Fleming's rules). The magnitude of the force experienced depends on the length of the conductor and on the strengths of the current and the magnetic field, and is greatest when the conductor is at right angles to the field. A conductor wound into a coil that can rotate between the poles of a magnet forms the basis of an ◊electric motor.

D

data facts, figures, and symbols, especially those stored in computers. The term is often used to mean raw, unprocessed facts, as distinct from information to which a meaning or interpretation has been applied.

decay, radioactive see ◊radioactive decay.

decibel unit (symbol dB) of measure, used originally to compare sound densities, and subsequently electrical or electronic power outputs; now also used to compare voltages. An increase of 10 dB is equivalent to a 10-fold increase in intensity or power, and a 20-fold increase in voltage. A whisper has an intensity of 20 dB; 140 dB (a jet aircraft taking off nearby) is the the threshold of pain.

declination, angle of see ◊angle of declination.

demagnetization the removal of magnetic properties from a material such as iron. An efficient method involves placing the specimen to be demagnetized inside a solenoid, or long coil of wire, carrying an alternating current, and gradually reducing that current to zero.

density measure of the compactness of a substance; it is equal to its mass per unit volume, and is measured in kilograms per cubic metre. The density D of a mass m kg occupying a volume V m^3 is given by the formula:

$$D = m/V$$

◊Relative density is the ratio of the density of a substance to that of water at 4°C.

diffraction the slight spreading of a light beam into a pattern of light and dark bands when it passes through a narrow slit or past the edge of an obstruction. The resulting patterns are known as interference phenomena. A **diffraction grating** is a device for separating a wave train such as a beam of incident light into its component frequencies (white light results in a spectrum).

digital in electronics, a term meaning 'coded as numbers'. A digital system uses two-state, either on/off or high/low voltage signals, to encode, receive, and transmit information. A *digital display*, such as that on a digital watch, shows discrete values as numbers (as opposed to an analogue display, such as the continuous sweep of the minute-hand on an analogue watch).

digital electronics is the technology that underlies digital techniques. Low-power, miniature, integrated circuits (chips) provide the means for the coding, storage, transmission, processing, and reconstruction of information of all kinds.

dimension basic physical quantity such as mass (M), length (L), and time (T), which can be combined by multiplication or division to give the dimensions of derived quantities. For example, acceleration (the rate of change of velocity) has dimensions (LT^{-2}), and is expressed in such units as km s^{-2}. A quantity that is a ratio, such as relative density or humidity, is dimensionless.

Dinorwig the location of Europe's largest pumped-storage hydroelectric scheme, completed 1984, in Gwynedd, North Wales. Six turbogenerators are involved, with a maximum output of some 1,880 megawatts. The working head of water for the station is 530 m.

The main machine hall is twice as long as a football field and as high as a 16-storey building.

diode a cold anode and a heated cathode (or the semiconductor equivalent, which incorporates a *p–n* junction). Either device allows the passage of direct current in one direction only, and so is commonly used in a ◊rectifier to convert alternating current (AC) to direct current (DC).

dioptre an optical unit in which the power of a ◊lens is expressed as the reciprocal of its focal length in metres. The usual convention is that convergent lenses are positive and divergent lenses negative. Short-sighted people need lenses of power about –0.66 dioptre; a typical value for long sight is about +1.5 dioptre.

dip, angle of see ♭angle of dip.

dipole, magnetic see ♭magnetic dipole.

direct current an electric current that flows in one direction, and does not reverse its flow as ♭alternating current does. The electricity produced by a battery is direct current.

dispersion in optics, the splitting of white light into a spectrum; for example, when it passes through a prism or a diffraction grating. It occurs because the prism (or grating) bends each component wavelength to a slightly different extent. The natural dispersion of light through raindrops creates a rainbow.

distance the extent of a journey or space between two points. It is a ♭scalar quantity, since it possesses magnitude but not direction. The SI unit of distance is the metre.

distance multiplier machine designed to convert a small movement into a large movement by multiplying its effect. It may also act as a *speed multiplier*. A fishing rod is a simple distance multiplier: a small movement of the hand holding the base of the rod produces a much larger movement at its tip. A simple speed multiplier is the pedal–chain–wheel mechanism of a bicycle in which slow rotation of the pedals produces a much faster movement of the rim of the wheels.

distance ratio alternative term for ♭velocity ratio.

distance–time graph graph used to describe the motion of a body by illustrating the relationship between the distance that it travels and the time taken. Plotting distance (on the vertical axis) against time (on the horizontal axis) produces a graph the gradient of which is the body's speed. If the gradient is constant (the graph is a straight line), the body has uniform or constant speed; if the gradient varies (the graph is curved), then so does the speed and the body may be said to be accelerating.

distributor device in a car engine's ignition system that distributes pulses of high-voltage electricity to the spark plugs in the cylinders. The electricity is passed to the plug leads by the tip of a rotor arm, driven by the engine camshaft, and current is fed to the rotor arm from the

distance-time graph

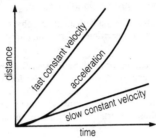

ignition coil. The distributor also houses the contact point or breaker, which opens and closes to interrupt the battery current to the coil, thus triggering the high-voltage pulses. In modern cars with electronic ignition, it is absent.

diverging lens lens that diverges or spreads out those light rays that have been refracted by it. It is a ◊concave lens, with one or both of its surfaces curving inwards. Such a lens is thinner at the centre than at the edge. Diverging lenses are used to correct ◊short-sightedness.

domain a small area in a magnetic material that behaves like a tiny magnet. Its magnetism is due to the movement of electrons in the atoms of the domain. In an unmagnetized sample, the domains point in random directions, or form closed loops, so that there is no overall magnetization of the sample. In a magnetized sample, the domains are aligned so that their magnetic effects combine to produce a strong overall magnetism.

Doppler effect change in observed frequency (or wavelength) of waves due to relative motion between wave source and observer. It is responsible for the perceived change in pitch of a siren as it approaches and then recedes, and for the ◊red shift of light from distant stars. It is named after the Austrian physicist Christian Doppler.

ductile material material that can sustain large deformations beyond its elastic limit (see ◊elasticity) without fracture. Metals are very ductile, and may be pulled out into wires, or hammered or rolled into thin sheets without breaking. Compare ◊brittle material.

dynamics in mechanics, the mathematical and physical study of the behaviour of bodies under the action of forces that produce changes of motion in them.

dynamo a simple generator, or machine for transforming mechanical energy into electrical energy. A dynamo in basic form consists of a powerful field magnet, between the poles of which a suitable conductor, usually in the form of a coil (armature), is rotated. The mechanical energy of rotation is thus converted into an electric current in the armature.

E

earth an electrical connection between an appliance and the ground. In the event of a fault in an electrical appliance, for example, involving connection between the live part of the circuit and the outer casing, the current flows to earth, causing no harm to the user.

In most domestic installations, earthing is achieved by a connection to a metal water-supply pipe buried in the ground before it enters the premises.

earth wire safety wire that connects the metal components of an electrical appliance to the earth or ground. It forms the third wire in a mains cable (the other two being the live and the neutral wires) and its insulator is usually coloured green with a yellow stripe.

If an appliance develops a fault so that current flows from the live wire to components that may be touched by the user, the earth wire will conduct that current to earth and prevent electric shock. The earth wire has a resistance lower than that of the live wire so that when a fault occurs, the current that now flows through both the earth and the live wires will increase to a level that will blow the ⊘fuse in the live wire or trip a ⊘circuit-breaker, thereby cutting off the electrical supply.

echo the repetition of a sound wave, or of a ⊘radar or ⊘sonar signal, by reflection from a surface. By accurately measuring the time taken for an echo to return to the transmitter, and by knowing the speed of a radar signal (the speed of light) or a sonar signal (the speed of sound in water), it is possible to calculate the range of the object causing the echo.

echo sounder device that detects objects under water by means of ⊘sonar—by using reflected sound waves. Most boats are equipped with echo sounders to measure the water depth beneath them. An echo sounder consists of a transmitter, which emits an ultrasonic pulse (see ⊘ultrasound), and a receiver, which detects the pulse after reflection from the seabed. The time between transmission and receipt of the reflected signal gives a measure of the depth of water.

eclipse the passage of an astronomical body through the shadow of another. The term is usually used for solar and lunar eclipses, which may be either partial or total, but also, for example, for eclipses by Jupiter of its satellites. An eclipse of a star by a body in the solar system is called an *occultation*.

A *solar eclipse* occurs when the Moon passes in front of the Sun as seen from Earth, and can happen only during a new Moon. During a *total eclipse* the Sun's corona can be seen. A total solar eclipse can last just over 7.5 minutes. When the Moon is at its farthest from Earth it does not completely cover the face of the Sun, leaving a ring of sunlight visible. This is an *annular eclipse* (from the Latin word *annulus* 'ring'). Between two and five solar eclipses occur each year.

A *lunar eclipse* occurs when the Moon passes into the shadow of the Earth, becoming dim until emerging from the shadow. Lunar eclipses may be partial or total, and they can happen only at full Moon. Total lunar eclipses last for up to 100 minutes; the maximum number each year is three.

eddy current an electric current induced in a conductor located in a changing magnetic field. Eddy currents can cause much wasted energy in the cores of transformers and other electrical machines.

efficiency in a machine, the useful work output (work done by the machine) divided by the work input (work put into the machine), usually expressed as a percentage. In formula terms:

$$\text{efficiency} = \frac{\text{useful work output}}{\text{work input}} \times 100\%$$

or, because power is the rate at which work is done:

$$\text{efficiency} = \frac{\text{useful power output}}{\text{power input}} \times 100\%$$

Losses of energy caused by friction mean that efficiency is always less than 100%, although it can approach this for electrical machines with no moving parts (such as a transformer).

eclipse

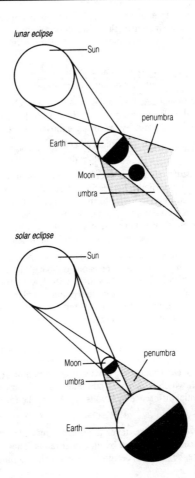

lunar eclipse

Sun — penumbra — Earth — Moon — umbra

solar eclipse

Sun — Moon — penumbra — umbra — Earth

Because work output may be defined as the product of the machine's load and the distance moved by that load, and work input as the product of the effort and the distance moved by the effort, efficiency may be also be expressed as:

$$\text{efficiency} = \frac{\text{load} \times \text{distance load is moved}}{\text{effort} \times \text{distance effort moves}} \times 100\%$$

or, because the ◊mechanical advantage (MA) of a machine is the ratio of the load to the effort, and its ◊velocity ratio (VR) is the distance moved by the effort divided by the distance moved by the load, as:

$$\text{efficiency} = \frac{\text{mechanical advantage}}{\text{velocity ratio}} \times 100\%$$

elastic collision collision between two or more bodies in which the total ◊kinetic energy of the bodies is conserved (remains constant); none is converted into any other form of energy. ◊Momentum also is conserved in such collisions. The molecules of a gas may be considered to collide elastically, but large objects may not because some of their kinetic energy will be converted on collision to heat and sound.

elasticity the ability of a solid to recover its shape once deforming forces (stresses modifying its dimensions or shape) are removed. An elastic material obeys ◊Hooke's law: that is, its deformation is proportional to the applied stress up to a certain point, called the *elastic limit*, beyond which additional stress will deform it permanently. Elastic materials include metals and rubber; however, all materials have some degree of elasticity.

electrical energy form of potential energy carried by an electric current. It may be converted into other forms of energy such as heat, light, and motion. The electrical energy W watts converted in a circuit component through which a charge Q coulombs passes and across which there is a potential difference of V volts is given by the formula:

$$W = QV$$

electrical safety measures taken to protect human beings from electric shock or from fires caused by electrical faults. They are of paramount importance in the design of electrical equipment. Safety measures include the fitting of earth wires and fuses or circuit breakers; the insulation of wires; the double insulation of portable equipment; and the use of residual-current devices (RCDs), which will break a circuit and cut off all currents if there is any imbalance between the currents in the live and neutral wires connected to an appliance (caused, for example, if some current is being conducted through a person).

The effects of electric shock vary from a tingling sensation to temporary paralysis and even death, and depend upon the amount of current passing through the body, and upon whether it passes through the central nervous system, thereby affecting brain and heart function. Fires are usually caused by overheated cables or loose connections.

electric bell a bell that makes use of electromagnetism. At its heart is a wire-wound coil on an iron core (an electromagnet) that, when a direct current (from a battery) flows through it, attracts an iron ◊armature. The armature acts as a switch, whose movement causes contact with an adjustable contact point to be broken, so breaking the circuit. A spring rapidly returns the armature to the contact point, once again closing the circuit, so bringing about the oscillation. The armature oscillates back and forth, and the clapper or hammer fixed to the armature strikes the bell.

electric charge property of some bodies that causes them to exert forces on each other. Two bodies both with positive or both with negative charges repel each other, whereas bodies with opposite or 'unlike' charges attract each other, since each is in the ◊electric field of the other. In atoms, ◊electrons possess a negative charge, and ◊protons an equal positive charge. The unit of electric charge is the coulomb (symbol C).

Atoms have no charge but can sometimes gain electrons to become negative *ions* or lose them to become positive ions. So-called ◊*static* electricity, seen in such phenomena as the charging of nylon shirts when they are pulled on or off, or in brushing hair, is in fact the gain or loss of electrons from the surface atoms. A flow of charge (such as electrons through a copper wire) constitutes an *electric current*; the flow of current is measured in *amperes* (symbol A).

electric field region in which a particle possessing electric charge experiences a force owing to the presence of another electric charge. It is a type of electromagnetic field.

electricity all phenomena caused by ◊electric charge, whether static or in motion. Electric charge is caused by an excess or deficit of electrons in the charged substance, and an electric current by the movement of electrons around a circuit. Substances may be electrical conductors, such as metals, which allow the passage of electricity through them, or insulators, such as rubber, which are extremely poor conductors. Substances with relatively poor conductivities that can be improved by the addition of heat or light are known as ◊semiconductors.

Electricity generated on a commercial scale was available from the early 1880s and used for electric motors driving all kinds of machinery, and for lighting, first by carbon arc, but later by incandescent filaments, first of carbon and then of tungsten, enclosed in glass bulbs partially filled with inert gas under vacuum. Light is also produced by passing electricity through a gas or metal vapour or a fluorescent lamp. Other practical applications include telephone, radio, television, X-ray machines, and many other applications in ◊electronics.

The fact that amber has the power, after being rubbed, of attracting light objects, such as bits of straw and feathers, is said to have been known to Thales of Miletus (around 600 BC) and to the Roman naturalist Pliny. William Gilbert, Queen Elizabeth I's physician, found that many substances possessed this power, and he called it 'electric' after the Greek word meaning 'amber'.

In the early 1700s, it was recognized that there are two types of electricity and that unlike kinds attract each other and like kinds repel. The charge on glass rubbed with silk came to be known as positive electricity, and the charge on amber rubbed with wool as negative electricity. These two charges were found to cancel each other when brought together.

In 1800 Alessandro Volta found that a series of cells containing brine, in which were dipped plates of zinc and copper, gave an electric current, which later in the same year was shown to evolve hydrogen and oxygen when passed through water (electrolysis). Humphry Davy, in 1807, decomposed soda and potash (both thought to be elements) and isolated the metals sodium and potassium, a discovery that led the way to electroplating. Other properties of electric currents discovered were the heating effect, now used in lighting and central heating, and the deflection of a magnetic needle, described by Hans Oersted 1820 and elaborated by André Ampère 1825. This work made possible the electric telegraph.

For Michael Faraday, the fact that an electric current passing through a wire caused a magnet to move suggested that moving a wire or coil of wire rapidly between the poles of a magnet would induce an electric current. He demonstrated this in 1831, producing the first ◊dynamo, which became the basis of electrical engineering. The characteristics of currents were formulated about 1827 by Georg Ohm, who showed that the current passing along a wire was equal to the electromotive force (emf) across the wire multiplied by a constant, which was the conductivity of the wire. The unit of resistance (ohm) is named after Ohm, the unit of emf is named after Volta (volt), and the unit of current after Ampère (amp).

The work of the late 1800s indicated the wide interconnections of electricity (with magnetism, heat, and light), and about 1855 James Clerk Maxwell formulated a single electromagnetic theory. The universal importance of electricity was decisively proved by the discovery that the atom, up until then thought to be the ultimate particle of matter, is composed of a positively charged central core, the nucleus, about which negatively charged electrons rotate in various orbits.

Electricity is the most useful and most convenient form of energy, readily convertible into heat and light and used to power machines. Electricity can be generated in one place and distributed anywhere because it readily flows through wires. It is generated at power stations where a suitable energy source is harnessed to drive ◊turbines that spin electricity generators. Common energy sources are coal, oil, water power (hydroelectricity), natural gas, and ◊nuclear energy. Research is under way to increase the contribution of wind, tidal, and geothermal power. Nuclear fuel has proved a more expensive source of electricity than initially anticipated and worldwide concern over radioactivity may limit its future development.

Electricity is generated at power stations at a voltage of about 25 kilovolts, which is not suitable for long-distance transmission. For minimal power loss, transmission must take place at very high voltage (400 kilovolts or more). The generated voltage is therefore increased ('stepped up') by a ◊transformer. The resulting high-voltage electricity is then fed into the main arteries of the ◊national grid system, an inter-connected network of power stations and distribution centres. After transmission to a local substation, the line voltage is reduced by a step-down transformer and distributed to consumers.

Among specialized power units that convert energy directly to electrical energy without the intervention of any moving mechanisms, the most promising are thermionic converters. These use conventional fuels such as propane gas, as in portable military power packs, or, if refuelling is to be avoided, radioactive fuels, as in uncrewed naviga-tional aids and spacecraft.

electric motor a machine that converts electrical energy into mechanical energy. There are various types, including direct-current and induction motors, most of which produce rotary motion. A linear induction motor produces linear (sideways) rather than rotary motion.

electrric motor

simple direct-current motor

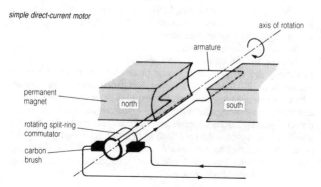

A simple ***direct-current motor*** consists of a horseshoe-shaped permanent ◊magnet with a wire-wound coil (◊armature) mounted so that it can rotate between the poles of the magnet.

A ◊commutator reverses the current (from a battery) fed to the coil on each half-turn, which rotates because of the mechanical force exerted on a conductor carrying a current in a magnetic field.

An ***induction motor*** employs ◊alternating current. It comprises a stationary current-carrying coil (stator) surrounding another coil (rotor), which rotates because of the current induced in it by the magnetic field created by the stator; it thus requires no commutator.

electrode conductor by which an electric current passes in or out of a substance.

electromagnet an iron bar with coils or wire around it, which acts as a magnet when an electric current flows through the wire. Electromagnets have many uses: in switches, electric bells, solenoids, and metal-lifting cranes.

electromagnetic field the agency by which a particle with an ◊electric charge experiences a force in a particular region of space. If it does so only when moving, it is in a pure ***magnetic field***; if it does so when stationary, it is in an ***electric field***. Both can be present simultaneously.

electromagnetic induction the production of an ◊electromotive force (emf) in a circuit by a change of magnetic flux through the circuit or by relative motion of the circuit and the magnetic flux. In a closed circuit an ◊induced current will be produced. All dynamos and generators make use of this effect. When magnetic tape is driven past the playback head (a small coil) of a tape-recorder, the moving magnetic field induces an emf in the head, which is then amplified to reproduce the recorded sounds.

electromagnetic radiation transfer of energy in the form of ◊electromagnetic waves.

electromagnetic spectrum the complete range, over all wavelengths from the lowest to the highest, of ◊electromagnetic waves.

electromagnetic waves oscillating electric and magnetic fields travelling together through space at a speed of nearly 300,000 km per second. The (limitless) range of possible wavelengths or ◊frequencies of electromagnetic waves, which can be thought of as making up the *electromagnetic spectrum*, includes radio waves, infrared radiation, visible light, ultraviolet radiation, X-rays, and gamma radiation.

electromotive force (emf) the energy supplied by a source of electric power in driving a unit charge around an electrical circuit. The unit is the ◊volt.

When the source is connected in circuit some of the energy it supplies will be lost in driving current across its own ◊internal resistance, and so its ◊terminal voltage (the potential difference across its terminals) will be less than its emf. If a source's terminal voltage is V volts, the current it supplies to a circuit is I amperes, and its internal resistance is r ohms, then its emf E can be expressed as:

$$E = V + Ir$$

or, where R is the total circuit resistance, as:

$$E = I(R + r)$$

electron stable, negatively charged elementary particle, a constituent of all ◊atoms and the basic particle of electricity. A beam of electrons will undergo ◊diffraction (scattering), and produce interference patterns, in the same way as ◊electromagnetic waves such as light; hence they may also be regarded as waves.

electron gun a part in many electronic devices consisting of a series of ◊electrodes, including a cathode for producing an electron beam. It plays an essential role in cathode-ray tubes (television tubes) and electron ◊microscopes.

electronics the branch of science that deals with the emission of ◊electrons from conductors and ◊semiconductors, with the subsequent manipulation of these electrons, and with the construction of electronic devices. The first electronic device was the thermionic valve, or vacuum tube, in which electrons moved in a vacuum, and led to such inventions as ◊radio, ◊television, ◊radar, and the digital ◊computer. Replacement of valves with the comparatively tiny and reliable transis-

electromagnetic waves

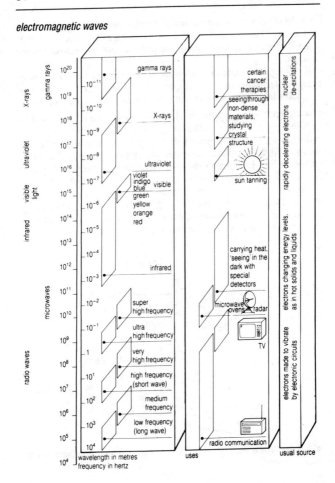

tor in 1948 revolutionized electronic development. Modern electronic devices are based on minute ◊integrated circuits (silicon chips), wafer-thin crystal slices holding tens of thousands of electronic components.

By using solid-state devices such as integrated circuits, extremely complex electronic circuits can be constructed, leading to digital watches, pocket calculators, powerful microcomputers, and word processors.

electroscope an apparatus for detecting ◊electric charge. The simple gold-leaf electroscope consists of a vertical conducting (metal) rod ending in a pair of rectangular pieces of gold foil, mounted inside and insulated from an earthed metal case. An electric charge applied to the end of the metal rod makes the gold leaves diverge, because they each receive a similar charge (positive or negative) and so repel each other.

The polarity of the charge can be found by bringing up another charge of known polarity and applying it to the metal rod. A like charge has no effect on the gold leaves, whereas an opposite charge neutralizes the charge on the leaves and causes them to collapse.

electroscope

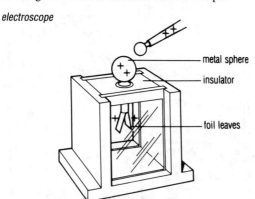

metal sphere

insulator

foil leaves

electrostatics the study of electric charges from stationary sources (not currents).

element substance that cannot be split chemically into simpler substances. The atoms of a particular element all have the same number of protons in their nuclei (their atomic number). Elements are classified in the periodic table. Of the 109 known elements, 95 are known to occur in nature (those with atomic numbers 1–95). Eighty-one of the elements are stable; all the others, which include atomic numbers 43, 61, and from 84 up, are radioactive. Those from 96 to 109 do not occur in nature and are synthesized only, produced in particle accelerators.

Elements are classified as metals, non-metals, or semimetals depending on a combination of their physical and chemical properties; about 75% are metallic. Some elements occur abundantly (oxygen, aluminium); others occur moderately or rarely (chromium, neon); some, in particular the radioactive ones, are found in minute (neptunium, plutonium) or very minute (technetium) amounts.

Symbols (devised by Jöns Berzelius) are used to denote the elements; the symbol is usually the first letter or letters of the English or Latinized name (for example, C for carbon, Ca for calcium, Fe for iron, *ferrum*). The symbol represents one atom of the element.

According to current theories, hydrogen and helium were produced in the 'Big Bang' at the beginning of the universe. Of the other elements, those up to atomic number 26 (iron) are made by nuclear fusion within the stars. The more massive elements such as lead and uranium, are produced when an old star explodes; as its centre collapses, the gravitational energy squashes nuclei together to make new elements.

emf abbreviation for ◊*electromotive force*.

energy the capacity for doing ◊work. *Potential energy* (PE) is energy deriving from position; thus a stretched spring has elastic PE; an object raised to a height above the Earth's surface, or the water in an elevated reservoir, has gravitational PE; a lump of coal and a tank of petrol, together with the oxygen needed for their combustion, have chemical PE (due to relative positions of atoms). Other sorts of PE include electrical and nuclear. Moving bodies possess *kinetic energy* (KE). Energy can be converted from one form to another, but the total quantity stays the same (in accordance with the conservation laws that govern many natural phenomena). For example, as an apple falls, it loses gravitational PE but gains KE.

So-called energy resources are stores of convertible energy. Nonrenewable resources include the fossil fuels (coal, oil, and gas) and ◊nuclear fission 'fuels' – for example, uranium-235. Renewable resources, such as wind, tidal, and geothermal power, have so far been less exploited. Hydroelectric projects are well established, and wind turbines and tidal systems are being developed. All energy sources depend ultimately on the Sun's energy.

Einstein's special theory of relativity 1905 correlates any gain, E, in KE with a loss, m, in 'rest mass', by the equation $E = mc^2$, in which E is energy and c is the speed of light. The equation applies universally, not just to nuclear reactions, although it is only for these that the percentage change in rest mass is large enough to detect. Although energy is never lost, after a number of conversions it tends to finish up as KE of random motion of molecules (of the air, for example) at relatively low temperatures. This is 'degraded' energy in that it is difficult to convert it back to other forms.

Burning fossil fuels causes acid rain and is gradually increasing the carbon dioxide content in the atmosphere, with unknown consequences for future generations. Coal-fired power stations also release significant amounts of radioactive material, and the potential dangers of nuclear power stations are greater still.

The ultimate nonrenewable but almost inexhaustible energy source would be nuclear fusion (the same way in which energy is generated in the Sun), but controlled fusion is a long way off. (The hydrogen bomb is a fusion bomb.) Harnessing resources generally implies converting their energy into electrical form, because electrical energy is easy to convert to other forms and to transmit from place to place, though not to store.

energy level or *shell* or *orbital* location of the electrons in an atom. The electrons in each level have a particular energy which is dependent upon the distance of that level from the nucleus of the atom. The levels are numbered beginning with one, the nearest to the nucleus. See ◊orbital, atomic.

energy use see under ◊alternative energy, ◊electricity, ◊gas, ◊green-house effect, ◊nuclear energy, ◊solar energy.

engine a device for converting stored energy into useful work or movement. Most engines use a fuel as their energy store. The fuel is burnt to produce heat energy—hence the name 'heat engine' – which is then converted into movement. Heat engines can be classifed according to the fuel they use (petrol engine or diesel engine), or according to whether the fuel is burnt inside (◊internal combustion engine) or outside (steam engine) the engine, or according to whether they produce a reciprocating or rotary motion (◊turbine or Wankel engine).

equations of motion mathematical equations that give the position or velocity at any time of an object moving with constant acceleration. The five common equations are:

$$v = u + at$$
$$s = \tfrac{1}{2}(u + v)t$$
$$s = ut + \tfrac{1}{2}at^2$$
$$s = vt + \tfrac{1}{2}at^2$$
$$v^2 = u^2 + 2as$$

in which a is the object's constant acceleration, u is its initial velocity, v is its velocity after a time t, and s is the distance travelled by it in that time.

equilibrium a steady state or condition of balance achieved by a body when the forces acting on it cancel each other (there is no resultant force), and the moments of those forces have no net result. In accordance with Newton's first law of motion, a body in equilibrium remains at rest or moves with constant velocity; it does not accelerate. See also ◊stability.

escape velocity minimum velocity with which an object must be projected for it to escape from the gravitational pull of a planetary body. In the case of the Earth, the escape velocity 11.2 kps.

evaporation process in which a liquid turns to a vapour without its temperature reaching boiling point. A liquid left to stand in a saucer eventually evaporates because, at any time, a proportion of its molecules will be fast enough (have enough kinetic energy) to escape from the attractive intermolecular forces at the liquid surface and into the atmosphere. The rate of evaporation rises with increased temperature because as the mean kinetic energy of the liquid's molecules rises so will the number possessing enough energy to escape.

A fall in the temperature of the liquid, known as the *cooling effect*, accompanies evaporation because as the faster proportion of the molecules escapes through the surface the mean energy of the remaining molecules falls. The effect may be noticed when wet clothes are worn, or as perspiration evaporates. ◊Refrigeration makes use of the cooling effect to extract heat from foodstuffs.

expansion the increase in size of a constant mass of substance (a body) caused by, for example, increasing its temperature (◊thermal expansion) or its internal pressure. The *expansivity*, or coefficient of thermal expansion, of a material is its expansion (per unit volume, area, or length) per degree rise in temperature.

eye the organ of vision. The *human eye* is a roughly spherical structure contained in a bony socket. Light enters it through the *cornea*, and passes through the circular opening (*pupil*) in the iris (the coloured part of the eye). The light is focused by the combined action of the curved cornea, the internal fluids, and the *lens* (the rounded transparent structure behind the iris). The ciliary muscles act on the lens to change its shape, so that images of objects at different distances can be focused on the *retina*. This is at the back of the eye, and is packed with light-sensitive cells (rods and cones), connected to the brain by the optic nerve. In contrast, the *insect eye* is compound—that is, made up of many separate facets, known as *ommatidia*, each of which collects light and directs it separately to a receptor to build up an image. Invertebrates, such as some worms and snails, and certain bivalves, have much simpler eyes, with no lens. Among molluscs, cephalopods have complex eyes similar to those of vertebrates. The mantis shrimp's eyes contain ten colour pigments with which to perceive colour; some flies and fishes have five, while the human eye has only three.

eye, defects of the abnormalities of the eye that impair vision. Glass or plastic lenses, in the form of spectacles or contact lenses, are the

usual means of correction. Common optical defects are ◊short-sighted-
ness or myopia; ◊long-sightedness or hypermetropia; lack of ◊accom-
modation or presbyopia; and astigmatism.

   Short-sightedness or myopia is the condition in which a person can
see clearly only those objects at distances of a few metres or less. It can
be corrected with a ◊diverging lens. In ***long sight*** or hypermetropia, a
person can see clearly only distant objects. This is corrected with a converging lens. In ***lack of accommodation*** or presbyopia, the eye's lens is
unable to adjust adequately in order to focus objects at different distances. This condition develops in almost all people's eyes from the
age of 40 onwards, and can be corrected only by using different lenses
for seeing short and long distances. ***Astigmatism*** is the condition in
which the curvature of the ◊cornea is uneven. It is corrected by means
of a cylindrical lens.

F

farad SI unit (symbol F) of electrical capacitance (how much electricity a ◊capacitor can store for a given voltage). One farad is a capacitance of one coulomb per volt. For practical purposes the microfarad (one millionth of a farad) is more commonly used.

far point the farthest point that a person can see clearly. The eye is unable to focus a sharp image on the retina of an object beyond this point. The far point for a normal eye should be at infinity; any eye that has a far point nearer than this is short-sighted (see ◊short-sightedness).

fast reactor or *fast breeder reactor* ◊nuclear reactor that makes use of fast neutrons to bring about fission. Unlike the ◊thermal reactor it makes little or no use of moderators to slow down neutrons. The reactor core is surrounded by a 'blanket' of uranium carbide. During operation, some of this uranium is converted into plutonium, which can be extracted and later used as fuel.

feedback general principle whereby the results produced in an ongoing reaction become factors in modifying or changing that reaction; the principle used in self-regulating control systems, from a simple thermostat to automatic computer-controlled machine tools. In such systems, information about what *is* happening in a system (such as level of temperature) is fed back to a controlling device, which compares it with what *should* be happening. If the two are different, the device takes suitable action (such as switching on a heater).

ferromagnetic material see ◊magnetic material.

fibre optics the branch of physics dealing with the transmission of light and images through glass or plastic fibres known as ◊optical fibres.

field the region of space in which an object exerts a force on another separate object because of certain properties they both possess. For example, there is a force of gravitational attraction between any two objects that have mass when one is in the gravitational field of the other.

Other fields of force include ◊electric fields (caused by electric charges) and ◊magnetic fields (caused by magnetic poles), either of which can involve attractive or repulsive forces.

field of view angle over which an image may be seen in a mirror or an optical instrument such as a telescope. A wide field of view allows a greater area to be surveyed without moving the instrument, but has the

field of view

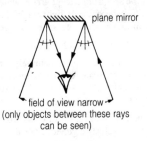

field of view narrow
(only objects between these rays
can be seen)

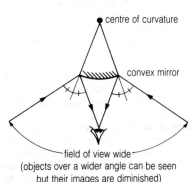

field of view wide
(objects over a wider angle can be seen
but their images are diminished)

disadvantage that each of the objects seen is smaller. A ◊convex mirror gives a larger field of view than a plane or flat mirror. The field of view of an eye is called its *field of vision* or visual field.

filter in electronics, a circuit that transmits a signal of some frequencies better than others. A low-pass filter transmits signals of low frequency and direct current; a high-pass filter transmits high-frequency signals; a band-pass filter transmits signals in a band of frequencies.

filter in optics, a device that absorbs some parts of the visible ◊spectrum and transmits others. For example, a green filter will absorb or block all colours of the spectrum except green, which it allows to pass through. A yellow filter absorbs only light at the blue and violet end of the spectrum, transmitting red, orange, green, and yellow light.

fission the splitting of the nucleus of an atom; see ◊nuclear fission.

fixed point a temperature that can be accurately reproduced and used as the basis of a temperature scale. In the Celsius scale, the fixed points are the temperature of melting ice, which is 0°C, and the temperature of boiling water (at standard atmospheric pressure), which is 100°C.

Fleming's rules memory aids for the directions of the magnetic field, current, and motion in an electric generator or motor, using one's fingers. The three directions are represented by the thu*m*b (for *m*otion), *f*orefinger (for *f*ield) and se*c*ond finger (*c*urrent), all held at right angles to each other. The right hand is used for generators and the left for motors. They were named after the English physicist John Fleming.

flip-flop in electronics, another term for a ◊bistable circuit.

floating state of equilibrium in which a body rests on or is suspended in the surface of a fluid (liquid or gas). According to ◊Archimedes' principle, a body wholly or partly immersed in a fluid will be subjected to an upward force, or upthrust, equal in magnitude to the weight of the fluid it has displaced. If the ◊density of the body is greater than that of the fluid, then its weight will be greater than the upthrust and it will sink. However, if the body's density is less than that of the fluid, the upthrust will be the greater and the body will be pushed upwards towards the surface. As the body rises above the surface the amount of

Fleming's rules

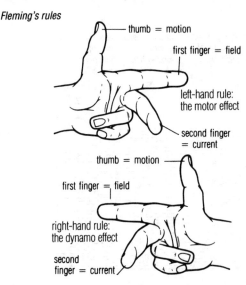

thumb = motion

first finger = field

left-hand rule:
the motor effect

second finger
= current

thumb = motion

first finger = field

right-hand rule:
the dynamo effect

second
finger = current

fluid that it displaces (and therefore the magnitude of the upthrust) decreases. Eventually the upthrust acting on the submerged part of the body will equal the body's weight, equilibrium will be reached, and the body will float.

flotation, law of law stating that a floating object displaces its own weight of the fluid in which it floats. It provides an explanation of how an object as large and as heavy as a steel ship can float: the hollow steel hull of the ship sinks into the water until the weight of the water it has displaced is as great as its own weight. The upthrust from the water will then equal the ship's weight and the ship will float.

fluid any substance, either liquid or gas, in which the molecules are relatively mobile and can 'flow'.

FM abbreviation for ◊frequency modulation.

floating

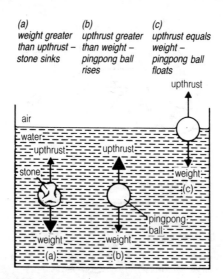

(a) weight greater than upthrust – stone sinks

(b) upthrust greater than weight – pingpong ball rises

(c) upthrust equals weight – pingpong ball floats

focal length the distance from the centre of a spherical mirror or lens to the focal point. For a concave mirror or convex lens, it is the distance at which parallel rays of light are brought to a focus to form a real image (for a mirror, this is half the radius of curvature). For a convex mirror or concave lens, it is the distance from the centre to the point at which a virtual image (an image produced by diverging rays of light) is formed.

In the case of lenses, the focal length is the reciprocal of the ◊power (in dioptres) of the lens: the greater its power, the shorter its focal length.

focus or *focal point* in optics, the point at which light rays converge, or from which they appear to diverge, to form a sharp image. Other electromagnetic rays, such as microwaves, and sound waves may also be brought together at a focus. Rays parallel to the principal axis of a

focal length

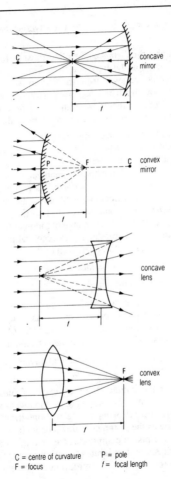

C = centre of curvature   P = pole
F = focus   *f* = focal length

lens or mirror are converged at, or appear to diverge from, the ◊principal focus.

focus in photography, the distance that a lens must be moved in order to focus a sharp image on the light-sensitive film at the back of the camera. The lens is moved away from the film to focus the image of closer objects. The focusing distance is often marked on a scale around the lens; however, some cameras have an automatic focusing (autofocus) mechanism that uses an electric motor to move the lens.

force any influence that tends to change the state of rest or the uniform motion in a straight line of a body. The action of an unbalanced or resultant force results in the acceleration of a body in the direction of action of the force or it may, if the body is unable to move freely, result in its deformation (see ◊Hooke's law). Force is a vector quantity, possessing both magnitude and direction; its SI unit is the newton.

According to Newton's second law of motion the magnitude of a resultant force is equal to the rate of change of ◊momentum of the body on which it acts; the force F producing an acceleration a m s^{-2} on a body of mass m kilograms is therefore given by:

$$F = ma$$

See also ◊Newton's laws of motion.

force multiplier machine designed to multiply a small effort in order to move a larger load. The number of times a machine multiplies the effort is called its ◊mechanical advantage. Examples of a force multiplier include a crowbar, wheelbarrow, nutcrackers, and bottle opener.

force ratio alternative term for ◊mechanical advantage, the ratio by which a force is magnified by a machine.

four-stroke cycle the engine-operating cycle of most petrol and diesel ◊engines. The 'stroke' is an upward or downward movement of a piston in a cylinder. In a petrol engine the cycle begins with the induction of a fuel mixture as the piston goes down on its first stroke. On the second stroke (up) the piston compresses the mixture in the top of the cylinder. An electric spark then ignites the mixture, and the gases produced force the piston down on its third, power stroke. On the fourth stroke (up) the piston expels the burned gases from the cylinder into the exhaust.

free fall the state in which a body is falling freely under the influence of gravity, as in free-fall parachuting. The term *weightless* is normally used to describe a body in free fall in space.

In orbit, astronauts and spacecraft are still held by gravity and are in fact falling towards the Earth. Because of their speed (orbital velocity), the amount they fall towards the Earth just equals the amount the Earth's surface curves away; in effect they remain at the same height, apparently weightless.

freezing change from liquid to solid state, as when water becomes ice. For a given substance, freezing occurs at a definite temperature, known as its freezing point, that is invariable under similar conditions of pressure. The temperature remains at this point until all the liquid is frozen. The amount of heat per unit mass that has to be removed to freeze a substance is a constant for any given substance, and is known as the latent heat of fusion.

frequency the number of periodic oscillations, vibrations, or waves occurring per unit of time. The unit of frequency is the hertz (Hz), one hertz being equivalent to one cycle per second. Human beings can hear sounds from objects vibrating in the range 20–15,000 Hz.

frequency modulation (FM) method of transmitting information, such as sound, over long distances by adding the information to a radio carrier wave. The frequency of the radio carrier wave increases when

frequency modulation

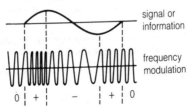

the signal amplitude increases, and decreases when the signal amplitude falls. FM radio broadcasts have better quality reception than amplitude-modulated or AM broadcasts (see ◊amplitude modulation) because the FM system is only concerned with changes in the frequency of the carrier wave, and is therefore not affected by the many forms of interference that change the amplitude of the carrier wave.

friction the force that opposes the relative motion of two bodies in contact. The *coefficient of friction* is the ratio of the force required to achieve this relative motion to the force pressing the two bodies together.

Friction is greatly reduced by the use of lubricants such as oil, grease, and graphite. Air bearings are now used to minimize friction in high-speed rotational machinery. In other instances friction is deliberately increased by making the surfaces rough—for example, brake linings, driving belts, soles of shoes, and tyres.

fuel any source of heat or energy, embracing the entire range of all combustibles and including anything that burns. *Nuclear fuel* is any material that produces energy in a nuclear reactor.

fundamental vibration standing wave of the longest wavelength that can be established on a vibrating object such as a stretched string or air column. The sound produced by the fundamental vibration is the lowest-pitched (usually dominant) note heard. The fundamental vibration of a string has a stationary ◊node at each end and a single ◊antinode at the centre where the amplitude of vibration is greatest.

fuse in electricity, a wire or strip of metal designed to melt when excessive current passes through. It is a safety device to stop at that point in the circuit surges of current that would otherwise damage equipment and cause fires.

fuse box insulated container housing the electrical fuses that protect the electric circuits and equipment in a building.

fusion the fusing together of atomic nuclei; see ◊nuclear fusion.

G

g symbol for the ◊gravitational field strength (the strength of the Earth's gravitational field at any point) and for gravitational acceleration.

gain in electronics, the ratio of the amplitude of the output signal produced by an amplifier to that of the input signal. In a ◊voltage amplifier the voltage gain is the ratio of the output voltage to the input voltage; in an inverting ◊operational amplifier (op-amp) it is equal to the ratio of the resistance of the feedback resistor to that of the input resistor.

galvanometer instrument for detecting small electric currents by their magnetic effect.

gamma radiation very high-frequency electromagnetic radiation emitted by the nuclei of radioactive substances during decay. The emission of gamma radiation reduces the energy of the source nucleus, but has no effect on its proton or nucleon numbers.

Rays of gamma radiation are stopped only by direct collision with an atom and are therefore very penetrating; they can, however, be stopped by about 4 cm of lead or by a very thick concrete shield. They are less ionizing in their effect than are alpha and beta particles, but are dangerous nevertheless because they can penetrate deeply into body tissues such as bone marrow. They are not deflected by either magnetic or electric fields.

Gamma radiation is used to kill bacteria and other microorganisms, sterilize medical devices, and change the molecular structure of plastics in order to modify their properties.

gas form of matter, such as air, in which the molecules move randomly in otherwise empty space, filling any size or shape of container into which the gas is put.

A sugar-lump sized cube of air at room temperature contains 30 million million million molecules moving at an average speed of 500 metres per second (1,800 kph). Gases can be liquefied by cooling, which lowers the speed of the molecules and enables attractive forces between them to bind them together.

gas-cooled reactor a type of nuclear reactor; see ◊advanced gas-cooled reactor (AGR).

gas laws physical laws concerning the behaviour of gases. They include ◊Boyle's law, ◊Charles's law, and the ◊pressure law, which are concerned with the relationships between the pressure, temperature, and volume of an ideal (hypothetical) gas.

These laws can be combined to give the *general* or *universal gas law*, which may be expressed as:

$$\frac{\text{pressure} \times \text{volume}}{\text{temperature}} = \text{constant}$$

or as:

$$\frac{P_1 V_1}{T_1} = \frac{P_2 V_2}{T_2}$$

gate see ◊logic gate.

gear a toothed wheel that transmits the turning movement of one shaft to another shaft. Gear wheels may be used in pairs, or in threes if both shafts are to turn in the same direction. The gear ratio, the ratio of the number of teeth on the two wheels, determines the torque ratio, the turning force on the output shaft compared with the turning shaft on the input shaft. The ratio of the angular velocities of the shafts is the inverse of the gear ratio.

Geiger counter device for detecting and/or counting nuclear radiation and particles. It detects the momentary current that passes between ◊electrodes in a suitable gas when a nuclear particle or a radiation pulse causes ionization in the gas. The electrodes are connected to electronic devices which enable the intensity of radiation or the number of particles passing to be measured. It is named after Hans Geiger.

generator a machine that produces electrical energy from mechanical energy, as opposed to an ◊electric motor, which does the opposite.

The ***dynamo*** is a simple generator consisting of a wire- wound coil (◊armature) which is rotated between the pole-pieces of a permanent magnet. The movement (of the wire in the magnetic field) induces a current in the coil by ◊electromagnetic induction, which can be fed by means of a ◊commutator as a continuous direct current into an external circuit. Slip rings instead of a commutator produce an alternating current, when the generator is called an ***alternator***.

gravitational field the region around an body in which other bodies experience a force due to its gravitational attraction. The gravitational field of a massive object such as the Earth is very strong and easily recognised as the force of gravity, whereas that of an object of much smaller mass is very weak and difficult to detect. Gravitational fields produce only attractive forces.

The ***gravitational force of attraction*** F between two masses m_1 and m_2 is given by Newton's universal law of gravitation:

$$F = \frac{Gm_1m_2}{r^2}$$

where G is the universal gravitational constant and r is the distance separating the centres of the two masses.

gravitational field strength (symbol g) the strength of the Earth's gravitational field at a particular point. It is defined as the the gravitational force in newtons that acts on a mass of one kilogram. The value of g on the Earth's surface is taken to be 9.806 N kg^{-1}.

The symbol g is also used to represent the acceleration of a freely falling object in the Earth's gravitational field. Near the Earth's surface and in the absence of friction due to the air, all objects fall with an acceleration of 9.806 m s^{-2}.

gravitational potential energy energy stored by an object when it is placed in a position from which, if it were free to do so, it would fall under the influence of gravity. The gravitational potential energy E_p of an object of mass m kilograms placed at a height h metres above the ground is given by the formula:

$$E_p = mgh$$

where $\lozenge g$ is the gravitational field strength in newtons per kilogram of the Earth at that place.

If a body possessing gravitational potential energy is released and allowed to fall, then that energy will be converted into $\lozenge$kinetic energy. The velocity v of the falling body that has fallen h metres may therefore be calculated by equating the formulae for gravitational potential energy and kinetic energy:

kinetic energy E_k = gravitational potential energy E_p

therefore:

$$\tfrac{1}{2}mv^2 = mgh$$

or, assuming that the mass of the body remains the same while falling:

$$\tfrac{1}{2}v^2 = gh$$

In a $\lozenge$hydroelectric power station, gravitational potential energy stored in water held in a high-level reservoir is used to drive turbines to produce electricity.

gravity the force of attraction between objects that exists because of their masses. The force we call gravity on Earth is the force of attraction between any object in the Earth's gravitational field and the Earth itself.

greenhouse effect a phenomenon of the Earth's atmosphere by which solar radiation, absorbed by the Earth and re-emitted from the surface, is prevented from escaping by gases, mainly carbon dioxide in the air. The result is a rise in the Earth's temperature; in a garden greenhouse, the glass walls have the same effect. The concentration of carbon dioxide in the atmosphere is estimated to have risen by 25% since the Industrial Revolution, and 10% since 1950; the rate of increase is now 0.5% a year. Other gases including chlorofluorocarbons (CFCs), methane, and nitrous oxide also play a part in the greenhouse effect.

The United Nations Environment Programme estimates an increase in average world temperatures of 1.5°C with a consequent rise of 20 cm in sea level by the year 2025.

grid the network by which electricity is generated and distributed over a region or country; see $\lozenge$national grid.

H

half-life during ◊radioactive decay, the time in which the strength of a radioactive source decays to half its original value. It may vary from millionths of a second to billions of years.

Radioactive substances decay exponentially; thus the time taken for the first 50% of the isotope to decay will be the same as the time taken by the next 25%, and by the 12.5% after that, and so on. For example, carbon-14 takes about 5,730 years for half the material to decay; another 5,730 for half of the remaining half to decay; then 5,730 years for half of that remaining half to decay, and so on. Plutonium-239, one of the most toxic of all radioactive substances, has a half-life of about 24,000 years. In theory, the decay process is never complete and there is always some residual radioactivity. For this reason, the half-life of a radioactive isotope is measured, rather than the total decay time.

heat a form of internal energy of a substance due to the kinetic energy of its molecules or atoms. Its SI unit is the joule. The extent to which a body will transfer or absorb heat (its hotness or coldness) is measured by ◊temperature, and is related to the mean kinetic energy of its molecules. Heat energy is transferred by ◊conduction, ◊convection current, and radiation (see ◊radiant heat), and always flows from a region of higher temperature to one of lower temperature. Its effect on a substance may be simply to raise its temperature, cause it to expand, melt it if a solid, vaporize it if a liquid, or increase its pressure if a confined gas.

The *specific heat* of a substance is the ratio of the quantity of heat required to raise the temperature of a given mass of the substance through a given range of temperature to the heat required to raise the temperature of an equal mass of water through the same range. It is measured by a calorimeter.

heat capacity alternative term for ◊thermal capacity.

heat pump machine, run by electricity or other power source, that cools the interior of a building by removing heat from interior air and pumping it out or, conversely, heats the inside by extracting energy from the atmosphere or from a hot-water source and pumping it in.

hertz SI unit (symbol Hz) of frequency (the number of repetitions of a regular occurrence in one second). Radio waves are often measured in megahertz (MHz), millions of hertz. It is named after Heinrich Hertz.

hi-fi (abbreviation of *hi*gh-*fi*delity) the faithful reproduction of sound from a machine that plays recorded music or speech. A typical hi-fi system includes a turntable for playing vinyl records, a cassette tape deck to play magnetic tape recordings, a tuner to pick up radio broadcasts, a compact-disc player, and an amplifier to serve all the equipment.

Hooke's law law stating that the deformation of a body is proportional to the magnitude of the deforming force, provided that the body's elastic limit (see ◊elasticity) is not exceeded. If the elastic limit is not reached, the body will return to its original size once the force is removed.

For example, if a spring is stretched by 2 cm by a weight of 1 N, it will be stretched by 4 cm by a weight of 2 N, and so on; however, once the load exeeds the elastic limit for the spring, Hooke's law will no longer be obeyed and each successive increase in weight will result in a greater extension until finally the spring breaks.

hydroelectric power electricity generated by moving water. In a typical hydroelectric power scheme, water stored in a reservoir, often created by damming a river, is piped into water ◊turbines, coupled to electricity generators. In ◊pumped storage plants, water flowing through the turbines is recycled. A tidal power station exploits the rise and fall of the tides. About one-fifth of the world's electricity comes from hydroelectric power.

hydrometer instrument used to measure the relative density of liquids (the density compared with that of water). A hydrometer consists of a thin glass tube ending in a sphere that leads into a smaller sphere, the latter being weighted so that the hydrometer floats upright, sinking deeper into lighter liquids than into heavier liquids. Hydrometers are used in brewing and to test the strength of acid in car batteries.

IC abbreviation for ◊integrated circuit.

image a picture or appearance of a real object, formed by light that passes through a lens or is reflected from a mirror. If rays of light actually pass through an image, it is called a ***real image***. Real images, such as those produced by a camera or projector lens, can be projected onto a screen. An image that cannot be projected onto a screen, such as that seen in a flat mirror, is known as a ***virtual image***.

impulse in mechanics, the product of a force and the time over which it acts. An impulse applied to a body causes its ◊momentum to change and is equal to that change in momentum. It is measured in newton seconds.

For example, the impulse J given to a football when it is kicked is given by:

$$J = Ft$$

where F is the kick force in newtons and t is the time in seconds for which the boot is in contact with the ball.

incident ray ray of light that strikes the surface of a mirror or meets a boundary between two different transparent materials. Such a ray arrives at an ◊angle of incidence to the normal to the surface or boundary.

inclined plane or ***ramp*** slope that allows a load to be raised gradually using a smaller effort than would be needed if it were lifted vertically upwards. It is a ◊force multiplier, possessing a ◊mechanical advantage greater than one. Bolts and screws are based on the principle of the inclined plane.

induced current electric current that appears in a closed circuit when there is relative movement of its conductor in a magnetic field. The

induced current

moving a magnet into a coil

coil of many turns
of insulated wire

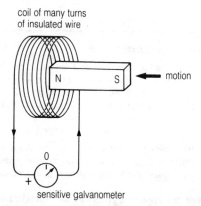

N S ← motion

0

+

sensitive galvanometer

effect is known as the ***dynamo effect***, and is used in all ⋄dynamos and generators to produce electricity. See ⋄electromagnetic induction.

There is no battery or other source of power in a circuit in which an induced current appears: the energy supply is provided by the relative motion of the conductor and the magnetic field. The magnitude of the induced current depends upon the rate at which the magnetic flux is cut by the conductor, and its direction is given by Fleming's right-hand rule (see ⋄Fleming's rules).

induction alteration in the physical properties of a body that is brought about by the influence of a field. See ⋄electromagnetic induction and ⋄magnetic induction.

inertia the tendency of an object to remain in a state of rest or uniform motion until an external force is applied, as stated by Newton's first law of motion (see ⋄Newton's laws of motion).

infrared radiation invisible electromagnetic radiation of wavelength between about 0.75 micrometres and 1 millimetre, that is, between the limit of the red end of the visible spectrum and the shortest microwaves. All bodies above the ◊absolute zero of temperature absorb and radiate infrared radiation. Infrared radiation is used in medical photography and treatment, and in industry, astronomy, and criminology.

input device a device for entering information into a computer. Input devices include keyboards, joysticks, touch-sensitive screens, graphics tablets, speech recognition devices, and vision systems.

insulation process or material that prevents or reduces the flow of electricity, heat, or sound from one place to another.
electrical insulation makes use of materials such as rubber, PVC, and porcelain, which do not conduct electricity, to prevent a current from leaking from one conductor to another or down to the ground. Insulation is a vital safety measure that prevents electric currents from being conducted through people and causing electric shock. *Double insulation* is a method of constructing electrical appliances that provides extra protection from electric shock, and renders the use of an earth wire unnecessary. In addition to the usual cable insulation, an appliance that meets the double insulation standard is totally enclosed in an insulating plastic body or structure so that there is no direct connection between any external metal parts and the internal electrical components.
thermal or *heat insulation* makes use of insulating materials such as fibreglass to reduce the loss of heat through the roof and walls of buildings. The ◊U-value of a material is a measure of its ability to conduct heat – a material chosen as an insulator should therefore have a low U-value. Air trapped between the fibres of clothes acts as a thermal insulator, preventing loss of body warmth.

insulator any poor ◊conductor of heat, sound, or electricity. Most substances lacking free (mobile) ◊electrons, such as non-metals, are electrical or thermal insulators.

integrated circuit (IC), popularly called *silicon chip* a miniaturized electronic circuit produced on a single crystal, or chip, of a semiconducting material such as silicon. It may contain many thousands of components and yet measure only 5 mm square and 1 mm thick. The IC

integrated circuit

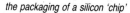

the packaging of a silicon 'chip'

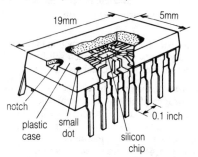

is held in a plastic or ceramic case, and linked via gold wires to metal pins with which it is connected to a ◊printed circuit board and the other components of devices such as computers and calculators.

intensity the power (or energy per second) per unit area carried by a form of radiation or wave motion. It is an indication of the concentration of energy present and, if measured at varying distances from the source, of the effect of distance on this. For example, the intensity of light is a measure of its brightness, and may be shown to diminish with distance from its source in accordance with the ◊inverse square law (intensity is inversely proportional to the square of the distance).

interference the phenomenon of two or more wave motions interacting and combining to produce a resultant wave of larger or smaller amplitude (depending on whether the combining waves are in or out of ◊phase with each other).

Interference of white light (multiwavelength) results in spectral coloured fringes, for example, the iridescent colours of oil films seen on water or soap bubbles. Interference of sound waves of similar frequency produces the phenomenon of ◊beats, often used by musicians when tuning an instrument. With monochromatic light (of a single wavelength), interference produces patterns of light and dark bands.

This is the basis of holography, for example. Interferometry can also be applied to radio waves, and is a powerful tool in modern astronomy.

internal-combustion engine engine in which fuel is burned inside the engine itself, contrasting with an external combustion engine (such as the steam engine) in which fuel is burned in a separate unit. The diesel and petrol engines are both internal-combustion engines. Gas turbines, and jet and rocket engines are sometimes also considered to be internal-combustion engines because they burn their fuel inside their combustion chambers.

internal reflection the reflection of light from the inside surface of a transparent material such as glass or water. When a light ray travelling through a dense material reaches the boundary between that material and a less dense medium such as air, some will pass through and be ◊refracted, but some will also be reflected back into the material. If the ◊angle of incidence exceeds a certain value, called the ◊critical angle for that material, ◊total internal reflection will occur and no light will escape.

internal resistance or *source resistance* the resistance inside a power supply, such as a battery of cells, that limits the current that it can supply to a circuit. For example, in order to supply the high current (hundreds of amperes) necessary to work the starter motor in a car, a battery with a very low internal resistance is required. One effect of internal resistance is to cause a drop in the ◊terminal voltage of a supply (the potential difference across its terminals) when that supply is connected in circuit; the difference between its terminal voltage and its ◊electromotive force (emf) increases as current flow increases (see ◊Ohm's law).

inverse square law the statement that the magnitude of an effect (usually a force) at a point is inversely proportional to the square of the distance between that point and the point location of its cause.

Light, sound, electrostatic force, gravitational force (Newton's law) and magnetic force (see ◊magnetism) all obey the inverse square law.

ion an atom, or group of atoms, which is either positively charged (*cation*) or negatively charged (*anion*), as a result of the loss or gain of electrons during chemical reactions or exposure to certain forms of radiation.

ionizing radiation radiation that knocks electrons from atoms during its passage, thereby leaving ions in its path. Such radiation is damaging to biological tissue. Alpha and beta particles are far more ionizing in their effect than are neutrons or gamma radiation.

isotope one of two or more atoms that have the same proton (atomic) number, but which contain a different number of neutrons, thus differing in their nucleon numbers. They may be stable or radioactive, naturally occurring or synthetic.

J

joule SI unit (symbol J) of work and energy. It is defined as the work done (energy transferred) by a force of one newton acting over one metre. It can also be expressed as the work done in one second by a current of one ampere at a potential difference of one volt. One watt is equal to one joule per second.

K

kelvin scale temperature scale used by scientists. It begins at ◊absolute zero (−273.16°C) and increases by the same degree intervals as the Celsius scale; that is, 0°C is the same as 273 K and 100°C is 373 K.

kilowatt-hour commercial unit of electrical energy (symbol kWh), defined as the work done by a power of 1,000 watts in one hour. It is used to calculate the cost of electrical energy taken from the ◊mains electric supply.

kinetic energy a form of ◊energy possessed by moving bodies. It is contrasted with ◊potential energy.

The kinetic energy of a moving body is equal to the work that would have to be done in bringing that body to rest, and is dependent upon both the body's mass and speed. The kinetic energy E_k (in joules) of a mass m kilograms travelling with speed v metres per second is given by the formula:

$$E_k = \frac{1}{2}mv^2$$

All atoms and molecules possess a certain amount of kinetic energy because they are all in some state of motion (see ◊kinetic theory). Adding heat energy to a substance increases the mean kinetic energy and hence the mean speed of its constituent molecules—a change that is reflected as a rise in the temperature of that substance.

kinetic theory theory describing the physical properties of matter in terms of the behaviour – principally movement – of its component atoms or molecules. A gas consists of rapidly moving atoms or molecules and, according to kinetic theory, it is their continual impact on the walls of the containing vessel that accounts for the pressure of the gas.

The slowing of molecular motion as temperature falls, according to kinetic theory, accounts for the physical properties of liquids and solids, culminating in the concept of no molecular motion at ◊absolute zero (0 K/–273°C). By making various assumptions about the nature of gas molecules, it is possible to derive from the kinetic theory the various gas laws (such as Avogadro's law, ◊Boyle's law, ◊Charles's law, and the ◊pressure law).

Kirchhoff's first law law formulated by Gustav Kirchhoff 1845 that states that the total current entering a junction in a circuit must equal the total current leaving it. It is based on the law of conservation of electric charge: as no charge can be created or destroyed at a junction, the flow of charge (current) into and out of the junction must be the same.

L

lamp, electric device designed to convert electrical energy into light energy.

In a *filament lamp*, such as a light bulb, an electric current causes heating of a long thin coil of fine high-resistance wire enclosed at low pressure inside a glass bulb. In order to give out a good light the wire must glow white-hot and therefore must be made of a metal, such as tungsten, that has a high melting point. The efficiency of filament lamps is low because most of the electrical energy is converted to heat.

A *fluorescent light* uses an electrical discharge or spark inside a gas-filled tube to produce light. The inner surface of the tube is coated with a fluorescent material that converts the ultraviolet light generated by the discharge into visible light. Although a high voltage is needed to start the discharge, these lamps are far more efficient than filament lamps at producing light.

laser (acronym for *l*ight *a*mplification by *s*timulated *e*mission of *r*adiation) a device for producing a narrow beam of light, capable of travelling over vast distances without dispersion and of being focused to give enormous power intensities (10^8 watts per cm^2 for high-energy lasers). Uses of lasers include communications (laser beams can carry far more information than can radio waves), cutting, drilling, welding, satellite tracking, medical and biological research, and surgery.

latent heat the heat required to change the state of a substance (for example, from solid to liquid) without changing its temperature.

lateral inversion the reversal experienced by an image formed in a plane (flat) mirror. Although the image is the correct way up, its left and right sides are transposed. The impression that most people have of their own face is based on the image that they see in a mirror; however, this is quite different from the face that other people see – for example, a hair parting that appears to be on the left when viewed in a mirror is seen by everyone else to be on the right.

Police vehicles and ambulances often have laterally inverted warning signs painted on their fronts, which appear the correct way round to a driver looking through a rear-view mirror.

LCD abbreviation for ◊*liquid crystal display*.

LDR abbreviation for ◊*light-dependent resistor*.

lead–acid cell a type of ◊accumulator (storage battery).

LED abbreviation for ◊*light-emitting diode*.

lens in optics, a piece of a transparent material, such as glass, possessing two polished surfaces – one concave or convex, and the other plane, concave, or convex – to modify rays of light. A convex lens brings rays of light together; a concave lens makes the rays diverge. Lenses are essential to spectacles, microscopes, telescopes, cameras, and almost all optical instruments.

The image formed by a single lens suffers from several defects or aberrations, notably spherical aberration in which a straight line becomes a curved image, and chromatic aberration in which an image in white light tends to have coloured edges. Aberrations are corrected by the use of compound lenses, which are built up from two or more lenses of different refractive index.

Lenz's law law stating that the direction of an electromagnetically induced current (generated by moving a magnet near a wire or a wire in a magnetic field) will oppose the motion producing it.

It is named after the German physicist Heinrich Friedrich Lenz (1804–65), who announced it in 1833.

lever a simple machine consisting of a rigid rod pivoted at a fixed point called the fulcrum, used for shifting or raising a heavy load or applying force in a similar way. Levers are classified into orders according to where the effort is applied, and the load-moving force developed, in relation to the position of the fulcrum.

A *first-order* lever has the load and the effort on opposite sides of the fulcrum (for example, a see-saw or pair of scissors).

lens

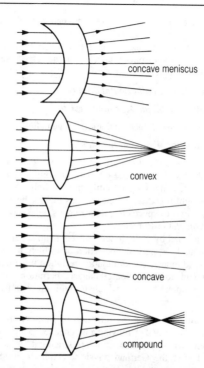

concave meniscus

convex

concave

compound

A *second-order* lever has the load and the effort on the same side with the load nearer the fulcrum (nutcrackers or a wheelbarrow).

A *third-order* lever has the effort nearer the fulcrum than the load with both on the same side of it (a pair of tweezers or tongs).

The ◊mechanical advantage of a lever is the ratio of load to effort, equal to the perpendicular distance of the effort's line of action from the fulcrum divided by the distance to the load's line of action. Thus tweezers, for instance, have a mechanical advantage of less than one.

lever

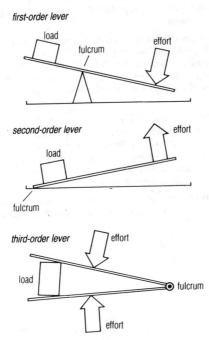

first-order lever

load

fulcrum

effort

second-order lever

load

effort

fulcrum

third-order lever

effort

load

fulcrum

effort

light ◊electromagnetic radiation in the visible range, having a wave-length from about 400 nanometres in the extreme violet to about 770 nanometres in the extreme red. Light is considered to exhibit both particle and wave properties, and the fundamental particle or quantum of light is called the *photon*. The speed of light (and of all electromagnetic radiation) in a vacuum is approximately 300,000 kms^{-1}. Newton was the first to discover, in 1666, that sunlight is composed of a mixture of light of different colours in certain proportions and that it could be separated into its components by dispersion. Before his time it was supposed that dispersion of light produced colour instead of separating already existing colours.

light bulb incandescent filament ◊lamp, first demonstrated by Joseph Swan in the UK 1878 and Thomas Edison in the USA 1879. The present-day light bulb is a thin glass bulb filled with an inert mixture of nitrogen and argon gas. It contains a filament made of fine tungsten wire. When electricity is passed through the wire, it glows white hot, producing light.

light-dependent resistor (LDR) component of electronic circuits whose resistance varies with the level of illumination on its surface. Usually resistance decreases as illumination rises. LDRs are used in light-measuring or light-sensing instruments (in the exposure-meter circuit of an automatic camera, for example) and in switches (such as those that switch on street lights at dusk).

light, deviation of the change in direction experienced by a light ray when it is reflected from a surface or when it is refracted at a surface

light, deviation of

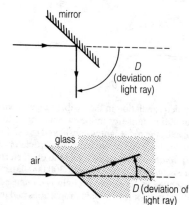

between two materials. The angle between the original (incident) ray and the reflected or refracted ray is the *angle of deviation*.

light-emitting diode (LED) means of displaying symbols in electronic instruments and devices. An LED is made of ◊semiconductor material, such as gallium arsenide phosphide, that glows when electricity is passed through it. The first digital watches and calculators had LED displays, but many later models use ◊liquid crystal displays.

lightning high-voltage electrical discharge between two charged rainclouds or between a cloud and the Earth, caused by the build-up of electrical charges. Air in the path of lightning ionizes (becomes conducting), and expands; the accompanying noise is heard as thunder. Currents of 20,000 amperes and temperatures of 30,000°C are common.

lightning conductor device that protects a tall building from lightning strike by providing an easier path for current to flow to earth than through the building. It consists of a thick copper strip of very low resistance that connects a metal rod projecting above the building to the ground below. A good connection to the ground is essential and is made by burying a large metal plate deep in the damp earth. In the event of a direct lightning strike, the current in the conductor may be so great as to melt or even vaporize the metal, but the damage to the building will nevertheless be limited.

line of force imaginary line representing the direction of a magnetic field. Its direction runs from the magnetic north pole to the magnetic south pole.

liquid state of matter between a ◊solid and a ◊gas. A liquid forms a level surface and assumes the shape of its container. Its atoms do not occupy fixed postions as in a crystalline solid, nor do they have freedom of movement as in a gas. Unlike a gas, a liquid is difficult to compress since pressure applied at one point is equally transmitted throughout. Hydraulics makes use of this property.

liquid crystal display (LCD) display of numbers (for example, in a calculator) or picture (such as on a pocket television screen) produced by molecules of a substance in a semiliquid state with some crystalline

liquid crystal display

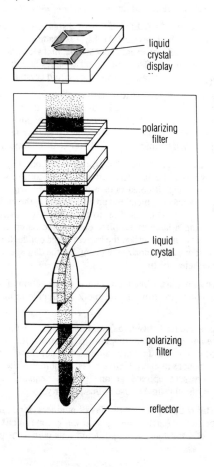

properties, in that clusters of molecules align in parallel formations. The display is a blank until the application of an electric field, which 'twists' the molecules so that they reflect or transmit light falling on them.

logic gate or *logic circuit* in electronics, one of the basic components used in building ◊integrated circuits. The five basic types of gate make logical decisions based on the functions NOT, AND, OR, NAND (NOT AND), and NOR (NOT OR). With the exception of the NOT gate, each has two or more inputs.

Information is fed to a gate in the form of binary-coded input signals (logic value 0 stands for 'off' or 'low-voltage pulse', logic 1 for 'on' or 'high-voltage'), and each combination of input signals yields a specific output (logic 0 or 1). An *OR* gate will give a logic 1 output if one or more of its inputs receives a logic 1 signal; however, an *AND* gate will yield a logic 1 output only if it receives a logic 1 signal through both its inputs. The output of a *NOT* or *inverter* gate is the opposite of the signal received through its single input, and a *NOR* or *NAND* gate produces an output signal that is the opposite of the signal that would have been produced by an OR or AND gate respectively.

logic gate

circuit symbols

| | OR gate | AND gate | NOT or inverter gate | NOR gate | NAND gate |

truth tables

inputs		output
0	0	0
0	1	1
1	0	1
1	1	1

OR gate

inputs		output
0	0	0
0	1	0
1	0	0
1	1	1

AND gate

inputs	output
0	1
1	0

NOT gate

inputs		output
0	0	1
0	1	0
1	0	0
1	1	0

NOR gate

inputs		output
0	0	1
0	1	1
1	0	1
1	1	0

NAND gate

The properties of a logic gate, or of a combination of gates, may be defined and presented in the form of a diagram called a ***truth table***, which lists the output that will be triggered by each of the possible combinations of input signals.

longitudinal wave wave in which the displacement of the medium's particles is in line with or parallel to the direction of travel of the wave motion. It is characterized by its alternating compressions and rarefactions. In the ***compressions*** the particles are pushed together (in a gas the pressure rises); in the ***rarefactions*** they are pulled apart (in a gas the pressure falls). Sound travels through air as a longitudinal wave.

longitudinal wave

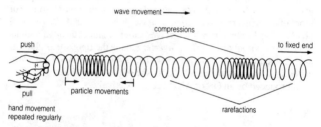

long-sightedness or ***hypermetropia*** defect of vision in which a person is able to focus on objects in the distance, but not on close objects. It is caused by the failure of the lens to return to its normal rounded shape, or by the eyeball being too short, with the result that the image is focused on a point behind the retina. Long-sightedness is corrected by wearing spectacles fitted with ◊converging lenses, each of which acts like a magnifying glass.

loudness subjective judgement of the level or power of sound reaching the ear. The human ear cannot give an absolute value to the loudness of a single sound, but can only make comparisons between two different sounds. Loudness is related to the amplitude of a sound wave and also, because the ear is not equally sensitive to all frequencies or pitches of sound, to frequency.

The precise measure of the power of a sound wave at a particular point is called its ◊intensity. Accurate comparisons of sound levels may be made using sound-level meters, which are calibrated in units called ◊decibels.

loudspeaker device that converts electrical signals into sound waves that are radiated into the air. It is used in all sound-reproducing systems such as radios, record players, tape recorders, and televisions.

The most common type is the *moving-coil speaker*. Electrical signals from, for example, a radio are fed to a coil of fine wire wound around the top of a cone. The coil is surrounded by a magnet. When signals pass through it, the coil becomes an electromagnet, which by moving causes the cone to vibrate, setting up sound waves.

loudspeaker

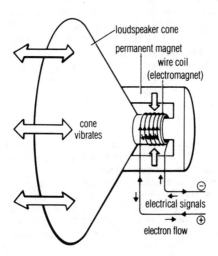

M

MA abbreviation for ◊mechanical advantage.

machine device that allows a small force (the effort) to overcome a larger one (the load). There are three basic machines: the sloping or ◊inclined plane, the ◊lever, and the ◊wheel and axle. All other machines are combinations of these three basic types. Simple machines derived from the inclined plane include the wedge and the screw; the spanner is derived from the lever; the pulley from the wheel.

The two principal features of a machine are its ◊mechanical advantage, which is the ratio load/effort, and its ◊efficiency, which is the work done by the load divided by the work done by the effort; the latter is expressed as a percentage. In a perfect machine, with no friction, the efficiency would be 100%. All practical machines have efficiencies of less than 100%, otherwise perpetual motion would be possible.

magnet any object that forms a magnetic field, either permanently or temporarily through induction, causing it to attract materials such as iron, cobalt, nickel, and alloys of these. It always has two ◊magnetic poles, called north and south.

magnetic compass device for determining the direction of the horizontal component of the Earth's magnetic field. It consists of a magnetized needle with its north-seeking pole clearly indicated, pivoted so that it can turn freely in a plane parallel to the surface of the Earth (in a horizontal circle). The needle will turn so that its north pole points towards the Earth's magnetic north pole.

Walkers, sailors, and other travellers use a magnetic compass to find their direction. The direction of the geographic, or true, North Pole is, however, slightly different from that of the magnetic north pole, and so the readings obtained from a compass of this sort must be adjusted using tables of magnetic corrections, or information marked on local maps.

magnetic dipole the pair of north and south magnetic poles, separated by a short distance, that makes up all magnets. Individual magnets are often called 'magnetic dipoles'. Single magnetic poles, or monopoles, have never been observed despite being searched for. See also magnetic ◊domain.

magnetic field the physical field or region around a permanent magnet or around a conductor carrying an electric current, in which a force acts on a moving charge or on a magnet placed in that field. The field can be represented by lines of force, which by convention link north and south poles and are parallel to the direction of a small compass needle placed on them. A magnetic field's magnitude and direction are given by the ◊magnetic flux density, expressed in teslas.

magnetic field

the Earth's magnetic field

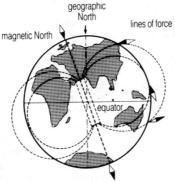

magnetic flux measurement of the strength of the magnetic field around electric currents and magnets. The amount of magnetic flux through an area equals the product of the area and the magnetic field strength at a point within that area. It is measured in webers; a density of one weber per square metre is equal to one tesla.

magnetic induction the production of magnetic properties in unmagnetized iron or other ferromagnetic material when it is brought close to a magnet. The material is influenced by the magnet's magnetic field and the two are attracted. The induced magnetism may be temporary, disappearing as soon as the magnet is removed, or permanent depending on the nature of the iron and the strength of the magnet. ◊Electromagnets make use of temporary induced magnetism to lift sheets of steel: the magnetism induced in the steel by the approach of the electromagnet enables it to be picked up and transported. To release the sheet, the current supplying the electromagnet is temporarily switched off and the induced magnetism disappears.

magnetic material or *ferromagnetic material* one of a number of substances that are strongly attracted by magnets and can be magnetized. These include iron, nickel, and cobalt, and all those ◊alloys that contain a proportion of these metals.

Soft magnetic materials can be magnetized very easily, but the magnetism induced in them (see ◊magnetic induction) is only temporary. They include Stalloy, an alloy of iron with 4% silicon used to make the cores of electromagnets and transformers, and the materials used to make 'iron' nails and paper clips.

Hard magnetic materials can be permanently magnetized by a strong magnetic field. Steel and special alloys such as Alcomax, Alnico, and Ticonal, which contain various amounts of aluminium, nickel, cobalt, and copper, are used to make permanent magnets. The strongest permanent magnets are ceramic, made under high pressure and at high temperature from powders of various metal oxides.

magnetic pole region of a magnet in which its magnetic properties are strongest. Every magnet has two poles, called north and south. The north (or north-seeking) pole is so named because a freely suspended magnet will turn so that this pole points towards the Earth's magnetic north pole. The north pole of one magnet will be attracted to the south pole of another, but will be repelled by its north pole. Like poles may therefore be said to attract, unlike poles to repel.

magnetism branch of physics dealing with the properties of magnets and ◊magnetic fields. Magnetic fields are produced by moving charged particles: in electromagnets, electrons flow through a coil of wire connected to a battery; in magnets, spinning electrons within the atoms generate the field.

magnification measure of the enlargement or reduction of an object in an imaging optical system. *Linear magnification* is the ratio of the size (height) of the image to that of the object. *Angular magnification* is the ratio of the angle subtended at the observer's eye by the image to the angle subtended by the object when viewed directly.

magnifying glass the simplest optical instrument, a hand-held converging lens used to produce a magnified, erect and virtual image. The image, being virtual, or an illusion created by the ◊refraction of light rays in the lens, can only be seen by looking through the magnifying glass.

The object to be magnified must be placed between the ◊principal focus and the lens; the image produced is best seen if it is at the ◊near point of the eye. The magnification produced by the lens is given by the ratio:

$$\text{magnification} = \text{height of image/height of object}$$

or, where the heights of the image and the object are h_I and h_O respectively:

$$\text{magnification} = h_I/h_O$$

magnifying glass

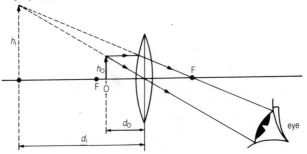

By similar triangles, this ratio can be shown to be equal to the distance ratio, therefore:

magnification = image distance/object distance

or, where the distances from the centre of the lens to the image and to the object are d_I and d_O respectively:

$$\text{magnification} = d_I/d_O$$

It follows that bringing the object closer to the eye increases the magnification.

mains electricity the domestic electricity-supply system. In the UK, electricity is supplied to houses, offices, and most factories as an ◊alternating current at a frequency of 50 hertz and a ◊root-mean- square voltage of 240 volts. An advantage of having an alternating supply is that it may easily be changed, using a ◊transformer, to a lower and safer voltage, such as 9 volts, for operating toys and for recharging batteries.

Maltese-cross tube cathode-ray tube used to demonstrate some of the properties of cathode rays. The cathode rays, or electron streams, emitted by the tube's ◊electron gun are directed towards a fluorescent screen in front of which hangs a metal Maltese cross. Those electrons

Maltese-cross tube

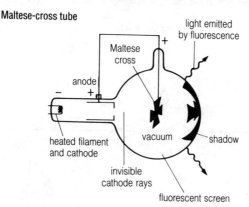

that hit the screen give up their kinetic energy and cause its phosphor coating to fluoresce. However, the sharply defined cross-shaped shadow cast on the screen shows that electrons are unable to pass through the Maltese cross. Cathode rays are thereby shown to travel in straight lines, and to be unable to pass through metal.

manometer instrument for measuring the pressure of liquids (including human blood pressure) or gases. In its basic form, it is a ◊U-tube partly filled with coloured liquid; the pressure of a gas entering at one side is measured by the level to which the liquid rises at the other.

mass the quantity of matter in a body as measured by its inertia. Mass determines the acceleration produced in a body by a given force acting on it, the acceleration being inversely proportional to the mass of the body. The mass also determines the force exerted on a body by ◊gravity on Earth, although this attraction varies slightly from place to place. In the SI system, the base unit of mass is the kilogram.

At a given place, equal masses experience equal gravitational forces, which are known as the weights of the bodies. Masses may, therefore, be compared by comparing the weights of bodies at the same place. The standard unit of mass to which all other masses are compared is a platinum-iridium cylinder of one kilogram.

mass number alternative name for the ◊nucleon number of an atom, the total number of neutrons and protons in the nucleus. ·

matter anything that has mass and can be detected and measured. All matter is made up of ◊atoms and elementary (or subatomic) particles, and exists ordinarily as a solid, liquid, or gas.

Maxwell's screw rule rule formulated by James Maxwell that predicts the direction of the magnetic field produced around a wire carrying electric current. It states that if a right-handed screw is turned so that it moves forwards in the same direction as the current, its direction of rotation will give the direction of the magnetic field.

mechanical advantage (MA) the number of times the load moved by a machine is greater than the effort applied to that machine. In equation terms:

$$MA = \text{load/effort}$$

melting point

98

Maxwell's screw rule

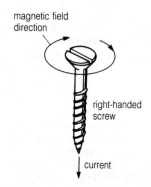

magnetic field
direction

right-handed
screw

current

MA has no units because it is a ratio. ◊Force multipliers have an MA
greater than one; ◊distance multipliers have an MA less than one.

The exact value of a working machine's MA is always less than its
predicted value because there will always be some frictional resistance
that increases the effort necessary to do the work.

melting point the temperature at which a substance melts, or changes
from solid to liquid form. A pure substance under standard conditions
of pressure has a definite melting point. If heat is supplied to a solid at
its melting point, the temperature does not change until the melting
process is complete. The melting point of ice is 0°C.

meniscus the curved shape of the surface of a liquid in a thin tube,
caused by the cohesive effects of ◊surface tension. Most liquids adopt a
concave curvature (viewed from above), although with highly viscous
liquids (such as mercury) the meniscus is convex. Meniscus is also the
name of a concavo-convex or convexo-concave ◊lens.

meter any instrument used for measurement; the term is often com-
pounded with a prefix to denote a specific type of meter; for example,
ammeter, voltmeter, or ratemeter.

meniscus

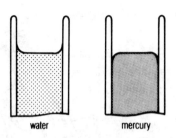

water   mercury

metric system system of weights and measures developed in France in the 18th century and recognized by other countries in the 19th century. In 1960 an international conference on weights and measures recommended the universal adoption of a revised International System (Système International d'Unités, or SI), with seven prescribed 'base units': the metre (m) for length, kilogram (kg) for mass, second (s) for time, ampere (A) for electric current, kelvin (K) for thermodynamic temperature, candela (cd) for luminous intensity, and mole (mol) for quantity of matter.

microphone the primary component in a sound-reproducing system, whereby the mechanical energy of sound waves is converted into electrical signals. One of the simplest is the telephone receiver mouthpiece, invented by Alexander Graham Bell in 1876; other types of microphone are used with broadcasting and sound-film apparatus.

microscope instrument for magnification with high resolution for detail. Optical and electron microscopes are the ones chiefly in use.

The *optical microscope* usually has two sets of glass lenses and an eyepiece. It was invented 1609 in the Netherlands by Zacharias Janssen.

The *electron microscope*, developed from 1932, passes a beam of electrons, instead of a beam of light, through a specimen and, since electrons are not visible, the eyepiece is replaced with a fluorescent screen or photographic plate; far higher magnification and resolution are possible than with the optical microscope.

microscope

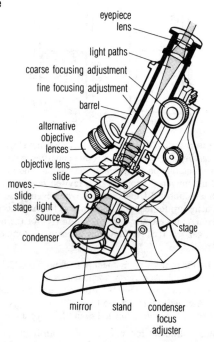

The *scanning electron microscope* (SEM), developed in the mid-1960s, moves a fine beam of electrons over the surface of a specimen, the reflected electrons being collected to form the image. The specimen has to be in a vacuum chamber.

microwave ◊electromagnetic wave with a wavelength in the range 0.3 to 30 cm, and a frequency of 300–300,000 megahertz (between radio waves and ◊infrared radiation). Microwaves are used in radar, in radio broadcasting, and in microwave heating and cooking.

microwave heating heating by means of microwaves. Microwave ovens use this form of heating for the rapid cooking or reheating of foods, where heat is generated throughout the food simultaneously. If food is not heated completely, there is a danger of bacterial growth that may lead to food poisoning. Industrially, microwave heating is used for destroying insects in grain and enzymes in processed food, pasteurizing and sterilizing liquids, and drying timber and paper.

mirage the illusion seen in hot climates of water on the horizon, or of distant objects being enlarged. The effect is caused by the ◊refraction of light.

Light rays from the sky bend as they pass through the hot layers of air near the ground, so that they appear to come from the horizon. Because the light is from a blue sky, the horizon appears blue and watery. If, during the night, cold air collects near the ground, light can be bent in the opposite direction, so that objects below the horizon appear to float above it. In the same way, objects such as trees or rocks near the horizon can appear enlarged.

mirror any polished surface that reflects light; often made from 'silvered' glass (in practice, a mercury alloy coating of glass). A plane (flat) mirror produces a same-size, erect, 'virtual' image located behind the mirror at the same distance from it as the object is in front of it. A spherical concave mirror produces a reduced, inverted, real image in front or an enlarged, erect, virtual image behind it (as with a shaving mirror), depending on how close the object is to the mirror. A spherical convex mirror produces a reduced, erect, virtual image behind it (as with a car's rear-view mirror).

moderator in a ◊thermal reactor, a material such as graphite or heavy water used to reduce the speed of high-energy neutrons.

modulation in radio transmission, the intermittent change of frequency, or phase amplitude, of a radio carrier wave, in accordance with the audio characteristics of the speaking voice, music, or other signal being transmitted. See ◊pulse-code modulation, ◊amplitude modulation (AM), and ◊frequency modulation (FM).

molecule the smallest unit of an ◊element or compound that can exist and still retain the characteristics of the element or compound. A molecule of an element consists of one or more similar ◊atoms; a molecule of a compound consists of two or more different atoms bonded together. Molecules vary in size and complexity from the hydrogen molecule (H_2) to the large macromolecules found in polymers.

moment of a force measure of the turning effect, or torque, produced by a force acting on a body. It is equal to the product of the force and the perpendicular distance from its line of action to the point, or pivot, about which the body will turn. Its unit is the newton metre.

If the magnitude of the force is F newtons and the perpendicular distance is d metres then the moment is given by:

$$\text{moment} = Fd$$

See also ◊couple.

momentum the product of the mass of a body and its velocity. If the mass of a body is m kilograms and its velocity is v metres per second, then its momentum is given by:

$$\text{momentum} = mv$$

Its unit is the kilogram metre-per-second (kg m s^{-1}) or the newton second.

The momentum of a body does not change unless a resultant or unbalanced force acts on that body (see ◊Newton's laws of motion). According to Newton's second law of motion, the magnitude of a resultant force F equals the rate of change of momentum brought about by its action, or:

$$F = (mv - mu)/t$$

where mu is the initial momentum of the body, mv is its final momentum, and t is the time in seconds over which the force acts. The change in momentum, or ◊impulse, produced can therefore be expressed as:

$$\text{impulse} = mv - mu = Ft$$

The law of ◊conservation of momentum is one of the fundamental concepts of classical physics. It states that the total momentum of all bodies in a closed system is constant and unaffected by processes occurring within the system.

motor anything that produces or imparts motion; a machine that provides mechanical power, particularly an ◊electric motor. Machines that burn fuel (petrol, diesel) are usually called ◊engines, but the internal-combustion engine that propels vehicles has long been called a motor, hence 'motorcar'. Strictly speaking, a car's motor is a part of its engine.

moving-coil meter instrument used to detect and measure electrical current. A coil of wire pivoted between the poles of a permanent magnet is turned by the motor effect of an electric current (by which a force acts on a wire carrying a current in a magnetic field). The extent to which the coil turns can then be related to the magnitude of the current. The sensitivity of the instrument depends directly upon the strength of the permanent magnet used, the number of turns making up the moving coil, and the coil's area. It depends inversely upon the strength of the controlling springs used to restrain the rotation of the coil. By the addition of a suitable resistor, such as a ◊multiplier, a moving-coil meter can be adapted to read potential difference in volts.

moving-coil meter

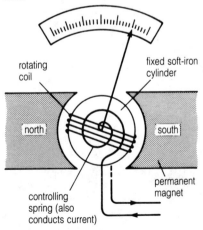

rotating coil

fixed soft-iron cylinder

north

south

permanent magnet

controlling spring (also conducts current)

multiplier resistor that converts a galvanometer or a current-detecting instrument such as a ⊅moving-coil meter into a voltage-reading volt-meter. It is always connected in series. The resistance of the multiplier selected reflects the range of voltage that the voltmeter will be required to measure.

The value of the resistor to be connected in series must increase the resistance of the galvanometer to a value which will limit the current through the instrument to that which produces a full-scale deflection when it reads the maximum voltage required.

N

NAND gate in electronics, a type of ◊logic gate.

national grid the network of cables, carried overhead on pylons or buried under the ground, that connects consumers of electrical power to power stations, and interconnects the power stations. It ensures that power can be made available to all customers at any time, allowing demand to be shared by several power stations, and particular power stations to be shut down for maintenance work from time to time. Britain has the world's largest grid system, with over 140 power stations able to supply up to 55,000 megawatts. See also ◊power transmission.

natural frequency the frequency at which a mechanical system will vibrate freely. A pendulum, for example, always oscillates at the same frequency when set in motion. This natural frequency depends upon the string's weight and tension.

More complicated systems, such as bridges, also vibrate with a fixed natural frequency. If a varying force with a frequency equal to the natural frequency is applied to such an object the vibrations can become violent, a phenomenon known as ◊resonance.

near point the closest position to the eye to which an object may be brought and still be seen clearly. For a normal human eye the near point is about 25 cm; however, it gradually moves further away with age, particularly after the age of 40.

neutral equilibrium the state of equilibrium possessed by a body that will stay at rest if moved into a new position; it will neither move back to its original position nor on any further. See ◊stability.

neutron one of the three chief subatomic particles (the others being the ◊proton and the ◊electron). Neutrons have about the same mass as protons but no electric charge, and occur in the nuclei of all ◊atoms except hydrogen. They contribute to the mass of atoms but do not affect their chemistry, which depends on the proton, or electron, numbers. For instance, ◊isotopes of a single element (with different masses) differ only in the number of neutrons in their nuclei and have identical chemical properties.

neutron number the number of neutrons possessed by an atomic nucleus. ◊Isotopes are atoms of the same element possessing different neutron numbers.

newton SI unit (symbol N) of ◊force. One newton is the force needed to accelerate an object with mass of one kilogram by one metre per second per second. To accelerate a car weighing 1,000 kg from 0 to 60 mph in 30 seconds would take about 250,000 N.

Newton's disc disc of card used to demonstrate that white light is composed of a spectrum of colours. It is divided into equal sectors, each of which is painted a different colour of the visible spectrum. If the disc is spun rapidly, its individual colours disappear and it appears off-white. Persistence of vision has caused the eye to 'remember' each of the colours it sees as they change places. The colours of the spectrum are thus added together giving rise to a colour that is almost white. Pure white is not seen because of the imperfect reflection of the colours of light from the disc.

Newton's laws of motion three laws that form the basis of Newtonian mechanics.

(1) Unless acted upon by an external resultant, or unbalanced, force, an object at rest stays at rest, and a moving object continues moving at the same speed in the same straight line. Put more simply, the law says that, if left alone, stationary objects will not move and moving objects will keep on moving at a constant speed in a straight line.

(2) A resultant or unbalanced force applied to an object produces a rate of change of ◊momentum that is directly proportional to the force and is in the direction of the force. For an object of constant mass m, this law may be rephrased as: a resultant or unbalanced force F applied to an object gives it an acceleration a that is directly proportional to, and in the direction of, the force applied and inversely proportional to the mass. This relationship is represented by the equation:

$$a = F/m$$

which is usually rearranged in the form:

$$F = ma$$

(3) When an object A applies a force to an object B, B applies an equal and opposite force to A; that is, to every action there is an equal and opposite reaction.

node a position in a ◊standing wave pattern at which there is no vibration. Points at which there is maximum vibration are called *antinodes*. Stretched strings, for example, can show nodes when they vibrate.

noise unwanted sound, especially one that is loud or disturbing. It is sometimes described as a form of pollution and this is particularly appropriate when the noise spoils another sound or makes that sound difficult to hear. At low levels, persistent noise can make people irritable, less alert, and less able to carry out skilled work accurately. At high levels it can cause temporary or permanent damage to hearing, nausea, and even temporaray blindness. Noise levels are measured in ◊decibels.

electronic noise takes the form of unwanted signals generated in electronic circuits and in recording processes by stray electrical or magnetic fields, or by temperature variations. In electronic recording and communication systems, 'white noise' frequently appears in the form of high frequencies, or hiss. The main advantages of digital systems are their relative freedom from such noise and their ability to recover and improve noise-affected signals.

NOR gate in electronics, a type of ◊logic gate.

NOT gate or *inverter gate* in electronics, a type of ◊logic gate.

nuclear energy energy from the inner core or ◊nucleus of the atom, as opposed to energy released in chemical processes, which is derived from the electrons surrounding the nucleus. See ◊nuclear reaction.

nuclear fission process whereby an atomic nucleus breaks up into two or more major fragments with the emission of several ◊neutrons.

nuclear fission

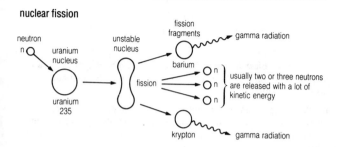

It is accompanied by the release of energy in the form of ¢gamma radiation and the ¢kinetic energy of the emitted particles.

Fission occurs spontaneously in nuclei of uranium-235, the main fuel used in nuclear reactors. However, the process can also be induced by bombarding nuclei with neutrons because a nucleus that has absorbed a neutron becomes unstable and soon splits. The neutrons released spontaneously by the fission of uranium nuclei may therefore be used in turn to induce further fissions, setting up a ¢chain reaction that must be controlled if it is not to result in a nuclear explosion.

nuclear fusion process whereby two atomic nuclei are 'melted' together, or fused, with the release of a large amount of energy. Very high temperatures and pressures are thought to be required in order for the process to happen. Under these conditions the atoms involved are stripped of all their electrons so that the remaining particles, which together make up a *plasma*, can come close together at very high speeds and overcome the mutual repulsion of the positive charges on the atomic nuclei. At very close range another nuclear force will come into play, fusing the particles together to form a larger nucleus. As fusion is accompanied by the release of large amounts of energy, the process might one day be harnessed to form the basis of commercial energy production. Methods of achieving controlled fusion are therefore the subject of research around the world.

Fusion is the process by which the Sun and the other stars produce their energy.

nuclear reaction reaction involving the nuclei of atoms. Atomic nuclei can undergo changes either as a result of radioactive decay, as in the decay of radium to radon (with the emission of an alpha particle) or as a result of being bombarded with particles.

◊Nuclear fission and ◊nuclear fusion are examples of nuclear reactions. The enormous amounts of energy released arise from the mass–energy relation put forward by Einstein, stating that $E = mc^2$ (where E is energy, m is mass, and c is the velocity of light).

In nuclear reactions the sum of the masses of all the products (on the atomic mass unit scale) is less than the sum of the masses of the reacting particles. This lost mass is converted to energy according to Einstein's equation.

nuclear reactor central component of a nuclear power station that generates ◊nuclear energy under controlled conditions for use as a source of electrical power. The nuclei of uranium-235 atoms undergo induced ◊nuclear fission in the reactor, and release energy in many forms, one of which is heat. The heat is removed from the core of the reactor by circulating gas or water, and is used to produce the steam that, under high pressure, drives turbines and alternators to produce electrical power. See also ◊thermal reactor, ◊advanced gas-cooled reactor (AGR), ◊pressurized water reactor (PWR), and ◊fast reactor.

nuclear waste the radioactive and toxic by-products of the nuclear-energy and nuclear-weapons industries. Reactor waste is of three types: high-level spent fuel, or the residue when nuclear fuel has been removed from a reactor and reprocessed; intermediate, which may be long- or short-lived; and low-level, but bulky, waste from reactors, which has only short-lived radioactivity. Disposal, by burial on land or at sea, has raised problems of safety, environmental pollution, and security. In absolute terms, nuclear waste cannot be safely relocated or disposed of.

nucleon any particle present in the atomic nucleus. ◊Protons and ◊neutrons are nucleons.

nucleon number or *mass number* the sum of the numbers of protons and neutrons in the nucleus of an atom. With the ◊proton number it is used in nuclear notation – for example, in the symbol $^{14}_{6}C$ representing the isotope carbon-14, the lower number is the proton number, and the upper is the nucleon number.

nucleus the positively charged central part of an ◊atom, which constitutes almost all its mass. Except for hydrogen nuclei, which have only ◊protons, nuclei are composed of both protons and ◊neutrons. Surrounding the nuclei are ◊electrons, of equal and opposite charge to that of the protons, thus giving the atom a neutral charge.

nuclide one of a species of atom possessing the same ◊proton number (atomic number) and the same ◊nucleon number (mass number).

O

ohm SI unit (symbol Ω) of electrical ◊resistance (the property of a substance that restricts the flow of electrons through it). It is defined as the resistance between two points when a potential difference of one volt between them produces a current of one ampere.

Ohm's law law that states that the current flowing in a metallic conductor maintained at constant temperature is directly proportional to the potential difference (voltage) between its ends. The law was discovered by Georg Ohm 1827.

If a current of I amperes flows between two points in a conductor across which the potential difference is V volts, then V/I is a constant called the ◊resistance R ohms between those two points. Hence:

$$V/I = R$$

or

$$V = IR$$

Not all conductors obey Ohm's law; those that do are called *ohmic conductors*.

operational amplifier (op-amp) a type of electronic circuit that is used to increase the size of an alternating voltage signal without distorting it.

Op-amps are used in a wide range of electronic measuring instruments. The name arose because they were originally designed to carry out mathematical operations and solve equations.

The voltage ◊gain of an inverting operational amplifier is equal to the ratio of the resistance of the feedback resistor to the resistance of the input resistor.

optical fibre very fine, optically pure, glass fibre through which light can be reflected to transmit an image or information from one end to the other. Bundles of such fibres are used in endoscopes to inspect otherwise inaccessible parts of machines or of the living body. Optical fibres are increasingly being used to replace copper wire in telephone cables, the messages being coded as pulses of light rather than a fluctuating electric current.

optical instrument instrument that makes use of one or more lenses or mirrors, or of a combination of these, in order to change the path of light rays and produce an image. Optical instruments such as magnifying glasses, microscopes, and telescopes are used to provide a clear, magnified image of the very small or the very distant. Others, such as cameras, photographic enlargers, and film projectors, may be used to store or reproduce images.

optics the branch of physics that deals with the study of light and vision—for example, shadows and mirror images, lenses, microscopes, telescopes, and cameras. For all practical purposes light rays travel in straight lines, although Einstein demonstrated that they may be 'bent' by a gravitational field. On striking a surface they are reflected or refracted with some absorption of energy, and the study of this is known as geometrical optics.

optoelectronics branch of electronics concerned with the development of devices (based on the ◊semiconductor gallium arsenide) that respond not only to the ◊electrons of electronic data transmission, but also to ◊photons.

In 1989, scientists at IBM in the USA built a gallium arsenide microprocessor ('chip') containing 8,000 transistors and four photodetectors. The densest optoelectronic chip yet produced, this can detect and process data at a speed of 1 billion bits per second.

orbital, atomic the region around the nucleus of an atom (or, in a molecule, around several nuclei) in which an ◊electron is most likely to be found. According to ◊quantum theory, the position of an electron is uncertain; it may be found at any point. However, it is more likely to be found in some places than in others, and it is these that make up the orbital.

OR gate in electronics, a type of ◊logic gate.

oscillation

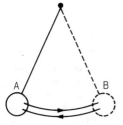

or

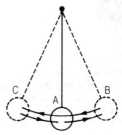

one complete oscillation
or cycle is from A to B
and back to A

one complete oscillation or cycle
is from A to B to C and back
to A, moving in the same
direction again

oscillation one complete to-and-fro movement of a vibrating object
or system. For any particular vibration, the time for one oscillation is
called its ◊period and the number of oscillations in one second is called
its ◊frequency. The maximum displacement of the vibrating object
from its rest position is called the ◊amplitude of the oscillation.

oscilloscope see ◊cathode-ray oscilloscope.

output device in computing, any device for displaying, in a form
intelligible to the user, the results of processing done by a computer.
The most common output devices are the VDU (visual display unit, or
screen) and the printer.

overtone a note that has a frequency or pitch that is a multiple of the
fundamental frequency, the sounding body's ◊natural frequency. Each
sound source produces a unique set of overtones, which gives the
source its quality or timbre.

P

parallel circuit electrical circuit in which current is split between two or more parallel paths or conductors. The division of the current across each conductor is in the ratio of their resistances. Thus, if the currents across two conductors of resistance R_1 and R_2, connected in parallel, are I_1 and I_2 respectively, then the ratio of those currents is given by the equation:

$$I_1/I_2 = R_2/R_1$$

The total resistance R of those conductors is given by:

$$1/R = 1/R_1 + 1/R_2$$

Compare ◊series circuit.

parallel circuit

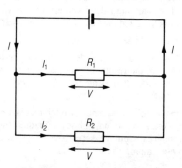

parallelogram of forces method of calculating the resultant (combined effect) of two different forces acting together on an object. Because a force has both magnitude and direction it is a ◊vector quantity and can be represented by a straight line. A second force acting at

parallelogram of forces

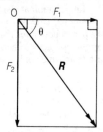

R is the resultant of F_1 and F_2

the same point in a different direction can be represented by another line drawn at an angle to the first. By completing the parallelogram (of which the two lines are sides) a diagonal may be drawn from the original angle to the opposite corner to represent the resultant force vector.

If two forces, F_1 and F_2, act at right-angles to each other on an object, the magnitude of their resultant R is given by the equation:

$$R^2 = F_1^2 + F_2^2$$

Its direction θ from F_1 is given by:

$$\tan \theta = F_2/F_1$$

pascal SI unit of pressure, equal to one newton per square metre.

PCM abbreviation for ◊pulse-code modulation.

pd abbreviation for ◊*potential difference*.

pendulum a weight (called a 'bob') swinging at the end of a cord or rod. When set in motion, it oscillates with a constant frequency that is inversely proportional to the length of its cord; each oscillation takes the same amount of time regardless of the size of its amplitude.

This regularity of swing was used in making the first accurate clocks in the 17th century. Pendulums can be used to measure the acceleration due to gravity, to measure velocities (ballistic pendulum), and to demonstrate the Earth's rotation (Foucault's pendulum).

penumbra the region of partial shade between the totally dark part (umbra) of a ◊shadow and the fully illuminated region outside. It occurs when a source of light is only partially obscured by a shadow-casting object. The darkness of a penumbra varies gradually from total darkness at one edge to full brightness at the other. In astronomy, a penumbra is a region of the Earth from which only a partial ◊eclipse of the Sun or Moon can be seen.

period time taken for one complete oscillation, or cycle of a repeated sequence of events. For example, the time taken for a pendulum to swing from side to side and back again is the period of the pendulum.

periscope optical instrument designed for observation from a concealed position such as from a submerged submarine. In its basic form it consists of a tube with parallel mirrors at each end, inclined at 45° to its axis.

phase stage in an oscillatory motion, such as a wave motion: two waves are in phase when their peaks and their troughs coincide. Otherwise, there is a *phase difference*, which has consequences in ◊interference phenomena and ◊alternating current electricity.

photocell or *photoelectric cell* a device for measuring or detecting light (or other electromagnetic radiation), since its electrical state is altered by the effect of light.

photodiode semiconductor *p–n* junction diode used to detect light or measure its intensity. The photodiode is encapsulated in a transparent plastic case that allows light to fall onto the junction. When this occurs, the reverse-bias resistance (high resistance in the opposite direction to normal current-flow) drops and allows a larger reverse-biased current to flow through the device. The increase in current can then be related to the amount of light falling on the junction. Photodiodes that can detect small changes in light level are used in alarm systems, camera exposure controls, and optical communication links.

photon the smallest 'package', 'particle', or quantum of energy in which ◊light, or any other form of electromagnetic radiation, is emitted. It has both particle and wave properties; it has no charge, is considered massless, but possesses momentum and energy.

physics the branch of science concerned with the ultimate laws that govern the structure of the universe, and the forms of matter and energy and their interactions. For convenience, physics is often divided into branches such as nuclear physics, electricity, electronics, magnetism, optics, and acoustics. Before this century, physics was known as *natural philosophy*.

pinhole camera the simplest type of camera, in which a pinhole rather than a lens is used to form an image. Light passes through the pinhole at one end of a box to form a sharp inverted image on the inside surface of the opposite end. The image is equally sharp for objects placed at different distances from the camera because only one ray from a particular distance or direction can enter through the tiny pinhole, and so only one corresponding point of light will be produced on the image. A photographic film or plate fitted inside the box will, if exposed for a long time, record the image.

pinhole camera

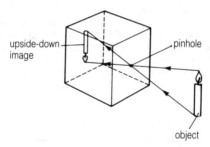

upside-down image — pinhole — object

pitch in mechanics, the distance between the adjacent threads of a screw or bolt. When a screw is turned through one full turn it moves up or down a distance equal to the pitch of its thread.

A screw thread is a simple type of machine, acting like a rolled-up ◊inclined plane, as may be illustrated by rolling up a paper ramp or long triangle around a pencil. A screw has a ◊mechanical advantage greater than one.

pitch

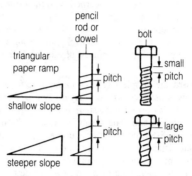

plug, three-pin insulated device with three metal projections used to connect the wires in the cable of an electrical appliance with the wires of a mains supply socket.

In the UK, plugs have pins of rectangular section, and must comply with the British Standard BS 1363 laid down by the British Standards Institute. A plug must be designed carefully with regard to construction, labelling, clearance between components, accessibility of live parts, earthing (see ▷earth), and terminal design. Before being approved, it is tested for the resistance of its insulation, temperature rise while in use, anchorage of cables, mechanical strength, and susceptibility to damage by heat and rust.

plutonium radioactive metallic element (symbol Pu) of proton number 94 and relative atomic mass 239.13. It occurs in nature in minute amounts, but large quantities of the isotope plutonium-239 are produced synthetically in nuclear reactors by bombarding uranium-238 with neutrons. Plutonium is one of the three elements capable of ▷nuclear fission (the others are thorium and uranium), and is used as a fuel in fast breeder reactors and in making nuclear weapons. It has a long half-life of 24,000 years and poses considerable disposal problems.

p–n junction diode in electronics, a two-terminal semiconductor device that allows electric current to flow in only one direction, the *forward-bias* direction. A very high resistance prevents current flow in the opposite, or *reverse-bias*, direction. It is used as a ◊rectifier, converting alternating current (AC) to direct current (DC).

The diode is cut from a single crystal of a semiconductor (such as silicon) to which special impurities have been added during manufacture so that the crystal is now composed of two distinct regions. One region contains semiconductor material of the p-type, which contains more positive charge carriers than negative; the other contains material of the n-type, which has more negative charge carriers than positive. The region of contact between the two types is called the p–n junction, and it is this that acts as the barrier preventing current from flowing, in conventional current terms, from the n-type to the p-type (in the reverse-bias direction).

pole, magnetic see ◊magnetic pole.

potential difference (pd) measure of the electrical potential energy converted to another form for every unit charge moving between two points in an electric circuit. In equation terms, potential difference V may be defined by:

$$V = W/Q$$

where W is the electrical energy converted in joules and Q is the unit charge in coulombs. The unit of potential difference is the volt. See also ◊Ohm's law.

potential divider or *voltage divider* two resistors connected in series in an electrical circuit in order to obtain a fraction of the potential difference, or input voltage, across the battery or electrical source. The potential difference is divided across the two resistors in direct proportion to their resistances. If the input voltage is V_{in} volts and the resistors have resistances of R_1 and R_2 ohms, then the output voltage V_{out}, or fraction of the potential difference across the first resistor, is given by the equation:

$$V_{out} = V_{in} \times R_1/(R_1 + R_2)$$

When a variable resistor, or *potentiometer*, is used as a potential divider, the output voltage can be varied continuously from zero to

potential divider

standard potential divider

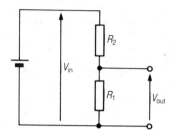

potentiometer used as a potential divider

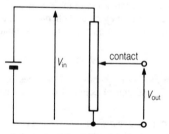

the value of the input voltage by sliding a contact along the resistor. Devices like this are used in electronic equipment as the variable controls for functions such as volume, tone, and brightness control.

potential, electric the relative electrical state of an object. A charged ◊conductor, for example, has a higher potential than the earth, whose potential is taken by convention to be zero. An electric ◊cell (battery) has a potential in relation to emf (◊electromotive force), which can make current flow in an external circuit. The difference in potential between two points – the potential difference – is expressed in ◊volts;

that is, a 12 V battery has a potential difference of 12 volts between its negative and positive terminals.

potential energy ◊energy possessed by an object by virtue of its relative position or state (for example, as in a compressed spring). It is contrasted with ◊kinetic energy. See ◊gravitational potential energy.

potentiometer a variable resistor that can be adjusted so as to compare, measure, or control voltages. A simple type consists of a length of uniform resistance wire (about 1 m long) carrying a constant current provided by a battery connected across the ends of the wire. The source whose potential difference (voltage) is to be measured is connected (to oppose the cell) between one end of the wire, through a galvanometer (instrument for measuring small currents), and a contact free to slide along the wire. The sliding contact is then moved until the galvanometer shows no deflection. The ratio of the length of potentiometer wire in the galvanometer circuit to the total length of wire is then equal to the ratio of the unknown potential difference to that of the battery. In radio circuits, any rotary variable resistance (such as volume control) is referred to as a potentiometer. See also ◊potential divider.

power in optics, a measure of the amount by which a lens will deviate light rays. A powerful converging lens will converge parallel rays steeply, bringing them to a focus at a short distance from the lens. The power of a lens, measured in dioptres, is equal to the reciprocal of its focal length in metres.

By convention, the power of a converging (or convex) lens is positive and that of a diverging (or concave) lens negative.

power the rate of doing work or consuming energy. Its SI unit is the watt (joule per second). If the work done or energy consumed is W joules and the time taken is t seconds, then the power P is given by the formula:

$$P = W/t$$

power, electric the rate at which an electrical machine or component converts ◊electrical energy into other forms of energy—for example, light, heat, or mechanical energy. Its SI unit is the watt (joule per second).

The power *P* of an electrical machine is given by the formula:

$$P = W/t$$

where *W* is the electrical energy converted by the machine over a time *t* seconds, or by the formulae:

$$P = IV$$

or

$$P = I^2R$$

where *I* is the current in amperes flowing through the machine, *V* is the potential difference in volts across it, and *R* is its resistance in ohms.

For example, an electric lamp that passes a current of 0.25 amperes at 240 volts uses 60 watts of electrical power and converts it into light – in everyday terms, it is a 60-watt lamp. An electric motor that requires 5 amperes at the same voltage consumes 1,200 watts (1.2 kilowatts).

power station building where electrical energy is generated from a fuel or from another form of energy. Fuels used include fossil fuels such as coal, gas, and oil, and the nuclear fuel uranium. Renewable sources of energy include ◊gravitational potential energy, used to produce ◊hydroelectric power, and ◊wind energy. The energy supply is used to turn ◊turbines either directly by means of water or wind pressure, or indirectly by steam pressure, steam being generated by burning fossil fuels or from the heat released by the nuclear fission of uranium nuclei. The turbines in their turn spin alternators, which generate electricity at very high voltage.

The largest power station in Europe is the Drax near Selby, Yorkshire, which supplies 10% of Britain's electricity.

power transmission transfer of electrical power from one location, such as a power station, to another . Electricity is conducted along the cables of the ◊national grid at a high voltage (up to 500 kV on the super grid) in order to reduce the current in the wires, and hence minimize the amount of energy wasted from them as heat. ◊Transformers are needed to step down these voltages before power can be supplied to consumers. High voltages require special insulators to prevent current from leaking to the ground and these may clearly be seen on the pylons that carry overhead wires.

pressure measure of the force acting normally (at right angles) to a body per unit surface area. The pressure p exerted by a force F newtons acting at right angles over an area of A square metres is given by:

$$p = F/A$$

The SI unit of pressure is the pascal (newton per square metre).

In a fluid (liquid or gas), atmospheric pressure increases with depth. At the edge of Earth's atmosphere, atmospheric pressure is zero, whereas at ground level it is about 100 kPa. The pressure p at a depth h in a fluid of density d is given by:

$$p = hdg$$

where g is the gravitational field strength. See also ◊U-tube.

pressure cooker closed pot in which food is cooked in water under pressure, where water boils at a higher temperature than normal boiling point (100°C) and therefore cooks food quickly. The modern pressure cooker has a quick-sealing lid and a safety valve that can be adjusted to vary the steam pressure inside.

pressure law law stating that the pressure of a fixed mass of gas at constant volume is directly proportional to its absolute temperature.

The law may be expressed as:

$$\text{pressure/temperature} = \text{constant}$$

or, more usefully, as:

$$P_1/T_1 = P_2/T_2$$

where P_1 and T_1 are the initial pressure and temperature in kelvin of a gas, and P_2 and T_2 are its final pressure and temperature. See also ◊gas laws.

pressurized water reactor (PWR) nuclear reactor used in many countries, and in nuclear-powered submarines. In the PWR, water under pressure is the coolant and ◊moderator. It circulates through a steam generator, where its heat boils water to provide steam to drive power ◊turbines.

principal focus in optics, the point at which incident rays parallel to the principal axis of a ◊lens converge, or appear to diverge, after refraction. The distance from the lens to its principal focus is its ◊focal length.

The principal focus of a converging (convex) lens or of a parabolic concave mirror is the point at which parallel incident rays will converge when refracted or reflected. It is a real focus.

The principal focus of a diverging (concave) lens is the point from which its parallel incident rays appear to diverge after refraction. It is a virtual or imaginary focus at which no light rays actually meet.

printed circuit board (PCB) an electrical circuit created by laying (printing) 'tracks' of a conductor such as copper onto one or both sides of an insulating board.

Components such as integrated circuits (silicon chips), resistors, and capacitors can be soldered to the surface of the board (surface mounted) or, more commonly, attached by inserting their connecting pins or wires into holes drilled in the board.

prism in optics, a triangular block of transparent material (plastic, glass, silica) commonly used to 'bend' a ray of light or split a beam into its spectral colours. Prisms are used as mirrors to define the optical path in binoculars, camera viewfinders, and periscopes. The dispersive property of prisms is used in the ◊spectrometer.

progressive wave or *travelling wave* wave that carries energy away from its source. The medium through which it moves does not in general travel with it. Every particle in the wave motion vibrates in the same way, but each vibration takes place slightly later than the vibration of the preceding particle. Although the shape of the wave stays the same as it travels, its amplitude gets smaller as its energy is absorbed by the medium or as the wave spreads out.

projectile particle that travels with both horizontal and vertical motion in the Earth's gravitational field. The two components of its motion can generally be analysed separately: its vertical motion will be accelerated due to its weight in the gravitational field; its horizontal motion may be assumed to be at constant velocity if the frictional forces of air resistance are ignored. In a uniform gravitational field and in the absence of frictional forces the path of a projectile is parabolic.

projector any apparatus that projects a picture on to a screen. In a *slide projector*, a lamp shines a light through the photographic slide or transparency, and a projection ◊lens throws an enlarged image of the slide onto the screen. A *film projector* has similar optics, but incorporates a mechanism that holds the film still while light is transmitted through each frame (picture). A shutter covers the film when it moves between frames.

proportion two variable quantities x and y are proportional if, for all values of x:

$$y = kx$$

where k is a constant. This means that if x increases, y increases in a linear fashion.

A graph of x against y would be a straight line passing through the origin (the point at which both x and y are equal to zero).

y is *inversely proportional* to x if the graph of y against $1/x$ is a straight line through the origin. The corresponding equation is:

$$y = k/x$$

Many laws of science relate quantities that are proportional (for example, ◊Boyle's law).

proton (Greek 'first') positively charged subatomic particle, a fundamental constituent of any atomic ◊nucleus. Its lifespan is effectively infinite.

A proton carries a unit positive charge equal to the negative charge of an ◊electron. Its mass is almost 1,836 times that of an electron, or 1.67×10^{-24} g. The number of protons in the atom of an ◊element is equal to its proton, or atomic, number.

proton number or *atomic number* the number of protons in the nucleus of an ◊atom. Adding the proton number to the number of neutrons in the nucleus (the neutron number) produces the ◊nucleon number.

pulley simple machine consisting of a fixed, grooved wheel, sometimes in a block, around which a rope or chain can be run. A simple pulley serves only to change the direction of the applied effort (as in a simple hoist for raising loads). The use of more than one pulley results in a ◊mechanical advantage, so that a given effort can raise a heavier load.

The mechanical advantage depends on the arrangement of the pulleys. For instance, a block and tackle arrangement with three ropes supporting the load will lift it with one-third of the effort needed to lift it directly (if friction is ignored), giving a mechanical advantage of 3.

pulse-code modulation (PCM) a form of digital ◊modulation in which microwaves or light waves (the carrier waves) are switched on and off in pulses of varying length according to a binary code. It is a relatively simple matter to transmit data that is already in binary code, such as that used by computer, by these means. However, if an analogue audio signal is to be transmitted, it must first be converted to a *pulse-amplitude modulated* signal (PAM) by regular sampling of its amplitude. The value of the amplitude is then converted into a binary code for transmission on the carrier wave.

pumped storage system in a hydroelectric plant that uses surplus electricity to pump water back into a high-level reservoir. In normal working the water flows from this reservoir through the ◊turbines to generate power for feeding into the grid. At times of low power demand, electricity is taken from the grid to turn the turbines into pumps that then pump the water back again. This ensures that there is always a maximum 'head' of water in the reservoir to give the maximum output when required.

PWR abbreviation for ◊pressurized water reactor.

Q

quantum theory the theory that many quantities, such as ◊energy, cannot have a continuous range of values, but only a number of discrete (particular) ones, because they are packaged in indivisible amounts called *quanta* (singular *quantum*). Just as earlier theory showed how light, generally seen as a wave motion, could also in some ways be seen as composed of particles (◊photons), quantum theory shows how atomic particles such as electrons may also be seen as having wavelike properties.

R

radar (acronym for *ra*dio *d*irection *a*nd *r*anging) device for locating objects in space, direction finding, and navigation by means of transmitted and reflected high-frequency radio waves. The direction of an object can be ascertained by transmitting a beam of short-wavelength (1–100 cm), short-pulse radio waves, and picking up the reflected beam. Distance is determined by timing the journey of the radio waves (travelling at the speed of light) to the object and back again.

Radar is essential to navigation in darkness, cloud, and fog, and is widely used in warfare to detect enemy aircraft and missiles. It may however be thwarted by modifying the shapes of aircraft and missiles in order to reduce their radar cross-section, and by means of devices such as radar-absorbent paints and electronic jamming. Radar is also used in meteorology and astronomy, and for detecting objects, such as service pipes or the remains of ancient buildings, underground.

radiant heat energy that is radiated by all warm or hot bodies. It belongs to the infrared part of the electromagnetic spectrum and causes heating when absorbed. Radiant heat is invisible and should not be confused with the red glow associated with very hot objects, which belongs to the visible part of the spectrum.

Infrared radiation can travel through a vacuum and it is in this form that the radiant heat of the Sun travels through space. It is the trapping of this radiation by carbon dioxide and methane in the atmosphere that gives rise to the ♭greenhouse effect.

radiation emission of radiant ♭energy as particles or waves—for example, heat, light, alpha particles, and beta particles (see ♭electromagnetic waves and ♭radioactivity).

radio transmission and reception of radio waves. In radio transmission a microphone converts ♭sound waves (pressure variations in the air) into ♭electromagnetic waves that are then picked up by a receiving aer-

radio

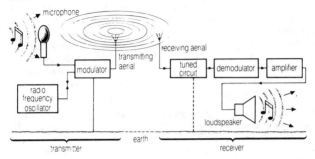

ial and fed to a loudspeaker, which converts them back into sound waves.

The theory of electromagnetic waves was first developed by James Clerk Maxwell 1864, given practical confirmation in the laboratory 1888 by Heinrich Hertz, and put to practical use by Marconi, who in 1901 achieved reception of a signal in Newfoundland transmitted from Cornwall, England. To carry the transmitted electrical signal, an oscillator produces a carrier wave of high frequency; different stations are allocated different transmitting carrier frequencies. A modulator superimposes the audiofrequency signal on the carrier. There are two main ways of doing this: amplitude modulation (AM), used for long- and medium-wave broadcasts, in which the strength of the carrier is made to fluctuate in time with the audio signal; and frequency modulation (FM), as used for VHF broadcasts, in which the frequency of the carrier is made to fluctuate. The transmitting aerial emits the modulated electromagnetic waves, which travel outwards from it.

In radio reception a receiving aerial picks up minute voltages in response to the waves sent out by a transmitter. A tuned circuit selects a particular frequency, usually by means of variable ◊capacitor connected across a coil of wire. A demodulator disentangles the audio signal from the carrier, which is now discarded, having served its purpose. An amplifier boosts the audio signal for feeding to the loudspeaker.

radioactive decay process of continuous disintegration undergone by the nuclei of radioactive elements, such as radium and uranium, and certain artificially created radioisotopes of stable elements. It results in the transformation of the original, or parent, nucleus into a daughter nucleus (which may have the proton number of an entirely different element, and may or may not be radioactive). Decay takes place at a specific rate expressed as the ◊half-life, which is the time taken for half of any mass of that particular isotope to decay completely. See also ◊alpha particle, ◊beta particle, and ◊gamma radiation.

radioactivity the spontaneous alteration, or decay, of the nuclei of radioactive atoms, accompanied by the emission of radiation. It is the property exhibited by the ◊radioisotopes (radioactive isotopes) of stable elements and by all the isotopes of radioactive elements, and can be either natural or induced.

Radioactivity establishes an equilibrium in parts of the nucleus of unstable radioactive substances, ultimately to form a stable arrangement of nucleons (protons and neutrons)—that is, a non-radioactive (stable) element. This is most frequently accomplished by the emission of ◊alpha particles (helium nuclei); ◊beta particles (electrons); or ◊gamma radiation (electromagnetic waves of very high frequency). It takes place either directly, through a one-step decay, or indirectly, through a number of decays that transmute one element into another. This is called a decay series or chain, and sometimes produces an element more reactive than its predecessor.

The instability of the particle arrangements in the nucleus of a radioactive atom (the ratio of neutrons to protons and/or the total number of both) determines the lengths of the ◊half-lives of the isotopes of that atom, which can range from fractions of a second to billions of years. Beta and gamma radiation are both ionizing and are therefore dangerous to body tissues, especially if a radioactive substance is ingested or inhaled.

radioactivity safety precautions taken to ensure the safe handling of radioactive materials. Such materials are hazardous because they emit ◊ionizing radiation, which damages living cells. The consequences of exposure to radiation or of contamination with radioactive materials—

for example, by inhaling or ingesting radioactive dust – may be immediate (burns, radiation sickness) or long-term (certain forms of cancer, birth defects due to genetic damage). Measures taken to protect people who work with radioactive sources include storing and transporting radioactive materials (including contaminated clothing and waste) in sealed, lead-lined containers; the use of thick leaded-glass, lead, or concrete barriers to shield workers from exposed materials; the use of tongs or remote-controlled devices; and the wearing of protective clothing and of monitoring devices, such as film badges, which keep a record of its wearer's exposure history.

radiocarbon dating or *carbon dating* method of dating organic materials (for example, bone or wood), used in archaeology. Plants take up carbon dioxide gas from the atmosphere and incorporate it into their tissues, and some of that carbon dioxide inevitably contains a certain amount of the radioactive isotope of carbon, carbon-14. On its death, the plant ceases to take up carbon-14 and the amount already taken up decays at a known rate, the half-life of 5,730 years, so that the time elapsed since the plant died can be measured in a laboratory. Animals take carbon-14 into their bodies from eating plant tissues and their remains can be similarly dated. After 120,000 years so little carbon-14 is left that no measure is possible.

radioisotope contraction of *radioactive isotope* naturally occurring or synthetic radioactive form of an element. Most radioisotopes are made by bombarding a stable element with neutrons in the core of a nuclear reactor. The radiations given off by radioisotopes are easy to detect (hence their use as tracers), can in some instances penetrate substantial thicknesses of materials, and have profound effects on living matter. Although dangerous, radioisotopes are used in the fields of medicine, industry, agriculture, and research. Most natural isotopes possessing a mass number below 208 are not radioactive.

radio telescope instrument for detecting and analysing radio waves emitted by sources such as galaxies and quasars in the universe. Radio telescopes usually consist of a steerable metal dish, up to 300 m across, that collects and focuses radio waves the way a concave mirror collects and focuses light waves. Other radio telescopes are shaped like long troughs, and some consist of simple rod-shaped aerials. Radio telescopes are much larger than optical telescopes, because the wavelengths they are detecting are much longer than the wavelength of light.

Radio interferometry is a technique in which the output from two or more dishes is combined to give better resolution of detail than with a single dish.

radio wave electromagnetic wave possessing a long wavelength (ranging from about 10^{-3} to 10^4 m) and a low frequency (from about 10^5 to 10^{11} Hz). Included in the radio-wave part of the spectrum are ◊microwaves, used for both communications and for cooking; ultra high and very high-frequency waves, which are used for television and FM (frequency-modulated) radio communications; and short, medium, and long waves, used for AM (amplitude-modulated) radio communications. Radio waves that are used for communications have all been modulated (see ◊modulation) to carry the information. Stars emit radio waves, which may be detected and studied using ◊radio telescopes.

rainbow arch in the sky displaying the colours of the ◊spectrum formed by the refraction and reflection of the Sun's rays through rain or mist. Its cause was discovered by Theodoric of Freiburg in the 14th century.

ramp another name for an ◊inclined plane, a slope used as a simple machine.

ratemeter instrument used to measure the ◊count rate of a radioactive source. It gives a reading of the number of particles emitted from the source and captured by a detector in a unit of time (usually a second).

ray diagram diagram that explains how and where mirrors and lenses form images. The paths of light rays are represented by lines, and standard constructions are used for certain rays that have known paths through a mirror or lens system. For example, a ray that arrives parallel to the principal axis of a converging lens will pass through its ◊principal focus on the far side of the lens; a ray passing through the centre of a lens is not deviated. If the paths of two rays from a single point on an object are plotted through a lens or mirror system, an image of that point will be formed where the paths of those two rays cross.

reaction force the equal and opposite force described by Newton's third law of motion that arises whenever one object applies a force (*action force*) to another. For example, if a magnet attracts a piece of iron, then that piece of iron will also attract the magnet with a force that is equal in magnitude but opposite in direction. When any object rests on the ground the downwards ◊contact force applied to the ground always produces an equal, upwards reaction force. See ◊Newton's laws of motion.

reaction principle principle stated by Isaac Newton as his third law of motion, namely, to every action there is an equal and opposite reaction. See ◊Newton's laws of motion.

reactor see ◊nuclear reactor.

receiver, radio component of a radio communication system that receives and processes radio waves. It detects and selects modulated radio waves (see ◊modulation) by means of an aerial and tuned circuit, and then separates the transmitted information from the carrier wave by a process that involves ◊rectification. The receiver device will usually also include the amplifiers that produce the audio signals (signals that are heard).

rectification process of converting ◊alternating current (AC) into ◊direct current (DC). It is necessary because almost all electrical power is generated, transmitted, and supplied as alternating current, but many devices, from television sets to electric motors, require direct current. It involves the use of one or more ◊diodes as rectifiers. However, a single diode produces half-wave rectification in which current flows in one direction for one-half of the alternating-current cycle only—an inefficient process in which the power is effectively turned off for half the time. A *bridge rectifier*, constructed from four diodes, can rectify the alternating supply in such a way that both the positive and negative halves of the alternating cycle can produce a current flowing in the same direction.

rectifier device used for obtaining one-directional current (DC) from an alternating source of supply (AC). Types include plate rectifiers, thermionic ◊diodes, and ◊semiconductor diodes.

red shift in astronomy, the lengthening of the wavelengths of light from an object as a result of the object's motion away from us. It is an example of the ◊Doppler effect. The red shift in light from galaxies is evidence that the universe is expanding.

reed switch switch containing two or three thin iron strips, or reeds, inside a sealed glass tube. When a magnet is brought near the switch, it induces magnetism in the reeds, which then attract or repel each other according to the positions of their induced magnetic poles. Electric current is switched on or off as the reeds make or break contact. A simple burglar alarm can be constructed by inserting a magnet into the closing edge of a door, and a reed switch into the part of the doorframe facing the magnet: when the door is opened the magnet moves away from the switch, closing the switch and setting off an alarm.

reflection the throwing back or deflection of waves, such as ◊light or ◊sound waves, when they hit a surface. The *law of reflection* states that

reflection

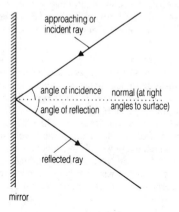

mirror

the angle of incidence (the angle between the ray and a perpendicular line drawn to the surface) is equal to the angle of reflection (the angle between the reflected ray and a perpendicular to the surface).

refraction the bending of a wave of light, heat, or sound when it passes from one medium to another. Refraction occurs because waves travel at different velocities in diferent media.

The refractive index of a material indicates by how much light is bent. See also ◊Snell's law of refraction, ◊apparent depth, and ◊internal reflection.

refraction

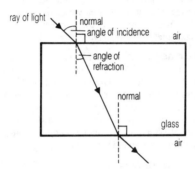

refractive index a measure of the refraction of a ray of light as it passes from one transparent medium to another. If the angle of incidence is i and the angle of refraction is r, the refractive index n is given by:

$$n = \sin i / \sin r$$

It is also equal to the speed of light in the first medium divided by the speed of light in the second, and it varies with the wavelength of the light.

refrigeration use of technology to transfer heat from cold to warm, against the normal temperature gradient, so that a body can remain substantially colder than its surroundings. Refrigeration equipment is used for the chilling and deep freezing of food, and in air conditioners and industrial processes.

Refrigeration is commonly achieved by a vapour-compression cycle, in which a suitable chemical (the refrigerant) travels through a long circuit of tubing, during which it changes from a vapour to a liquid and back again. A compression chamber makes it condense, and thus give out heat. In another part of the circuit, called the evaporator coils, the pressure is much lower, so the refrigerant evaporates, absorbing heat as it does so. The evaporation process takes place near the central part of the refrigerator, which therefore becomes colder, while the compression process takes place near a ventilation grille, transferring the heat to the air outside. The most commonly used refrigerants in modern systems were chlorofluorocarbons (CFCs), but these are now being replaced by coolants that do not damage the ozone layer.

relative atomic mass a measure of the mass of an atom. It depends on the number of protons and neutrons in the atom, the electrons having negligible mass. It is calculated relative to one-twelfth the mass of an atom of carbon-12. If more than one ◊isotope of the element is present, the relative atomic mass is calculated by taking an average that takes account of the relative proportions of each isotope, resulting in values that are not whole numbers. The term *atomic weight*, although commonly used, is strictly speaking incorrect.

relative density or *specific gravity* the density (at 20°C) of a solid or liquid relative to (divided by) the maximum density of water (at 4°C). The relative density of a gas is its density divided by the density of hydrogen (or sometimes dry air) at the same temperature and pressure.

relay in electrical engineering, an electromagnetic switch. A small current passing through a coil of wire wound around an iron core attracts an ◊armature whose movement closes a pair of sprung contacts to complete a secondary circuit, which may carry a large current or activate other devices. The solid-state equivalent is a thyristor switching device.

resistance the property of a substance that restricts the flow of electricity through it, associated with the conversion of electrical energy to heat; also the magnitude of this property. Resistance depends on many factors, such as the nature of the material, its temperature, dimensions, and thermal properties; degree of impurity; the nature and state of illumination of the surface; and the frequency and magnitude of the current. The SI unit of resistance is the ohm. See also ◊Ohm's law.

resistivity measure of the ability of a material to resist the flow of an electric current. It is numerically equal to the ◊resistance of a sample of unit length and unit cross-sectional area, and its unit is the ohm metre. If a conductor of length l metres and cross- sectional area A square metres has a resistance R ohms, then its resistivity ρ is given by:

$$\rho = RA/l$$

A good conductor has a low resistivity (1.7×10^{-8} ohm metres for copper); an insulator has a very high resistivity (10^{15} ohm metres for polyethene).

resistor any component in an electrical circuit used to introduce ◊resistance to a current. Resistors are often made from wire-wound coils or pieces of carbon. ◊Rheostats and ◊potentiometers are variable resistors.

resolution of forces in mechanics, the division of a single force into two parts that act at right angles to each other. The two parts of a resolved force, called its *components*, have exactly the same effect on an object as the single force which they replace.

For example, the weight W of an object on a slope, tilted at an angle θ, can be resolved into two components: one acting at a right angle to the slope, equal to $W\cos \theta$, and one acting parallel to and down the slope, equal to $W\sin \theta$. The component acting down the slope (minus any friction force that may be acting in the opposite direction) is responsible for the acceleration of the object.

resonance rapid and uncontrolled increase in the size of a vibration when the vibrating object is subject to a force varying at its ◊natural frequency. In a trombone, for example, the length of the air column in the instrument is adjusted until it resonates with the note being sounded. Resonance effects are also produced by many electrical circuits. Tuning a radio, for example, is done by adjusting the natural frequency

resolution of forces

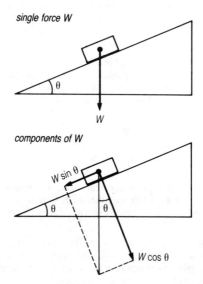

single force W

W

components of W

$W \sin \theta$

$W \cos \theta$

of the receiver circuit until it coincides with the frequency of the radio waves falling on the aerial.

Resonance has many physical applications. Children use it to increase the size of the movement on a swing, by giving a push at the same point during each swing. Soldiers marching across a bridge in step could cause the bridge to vibrate violently if the frequency of their steps coincided with its natural frequency. Resonance was the cause of the collapse of the Tacoma Narrows bridge, USA, in 1940 when the frequency of the wind coincided with the natural frequency of the bridge.

resultant force in mechanics, a single force acting on a particle or body whose effect is equivalent to the combined effects of two or more component forces. The resultant of two forces acting at one point on an object can be found using the ◊parallelogram of forces method.

retina the inner layer at the back of the vertebrate ◊eye, which contains light-sensitive cells and nerve fibres. Light falling on the retina produces chemical changes in the cells causing them to send electrical signals along the nerve fibres via the optic nerve to the brain. The light-sensitive cells are of two types, called rod cells and cone cells after the shape of their outer projections. The *rod cells*, about 120 million in each eye, are distributed throughout the retina. They are sensitive to low levels of light, but do not provide detailed or sharp images, nor can they detect colour. The *cone cells*, about 6 million in number, are mostly concentrated in a central region of the retina called the *fovea*, and provide both detailed and colour vision. The cones of the human eye contain three visual pigments, each of which responds to a different primary colour (red, green, or blue). The brain can interpret the varying signal levels from the three types of cone as any of the different colours of the visible spectrum.

reverberation in acoustics, the multiple reflections, or echoes, of sounds inside a building that merge and persist a short while (up to a few seconds) before fading away. At each reflection some of the sound energy is absorbed, causing the amplitude of the sound wave and the intensity of the sound to reduce a little. Too much reverberation causes sounds to become confused and indistinct, and this is particularly noticeable in empty rooms and halls, and buildings such as churches and cathedrals where the hard, unfurnished surfaces do not absorb sound energy well. Where walls and surfaces absorb sound energy very efficiently, too little reverberation may cause a room or hall to sound dull or 'dead'. Reverberation is a key factor in the design of theatres and concert halls, and can be controlled by lining ceilings and walls with materials possessing specific sound-absorbing properties.

rheostat variable ◊resistor, usually consisting of a high-resistance wire-wound coil with a sliding contact. It is used to vary electrical resistance without interrupting the current (for example, when dimming lights). The circular type in electronics (which can be used, for

rheostat

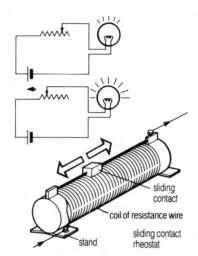

example, as the volume control of an amplifier) is also known as a ◊potentiometer.

ring circuit household electrical circuit in which appliances are connected in series to form a ring with each end of the ring connected to the power supply.

ripple tank shallow water-filled tray used to demonstrate various properties of waves, such as reflection, refraction, diffraction, and interference, by programming and manipulating their movement.

rocket projectile driven by the reaction of gases produced by a fast-burning fuel. Unlike jet engines, which are also reaction engines, modern rockets carry their own oxygen supply to burn their fuel and are totally independent of any surrounding atmosphere. As rockets are the only form of propulsion available that can function in a vacuum, they are essential to exploration in outer space. Multistage rockets have to be used, consisting of a number of rockets joined together.

root-mean-square current (rms current) the numerical value given to an ◊alternating current that converts exactly the same power in a resistor as a ◊direct current of the same steady value. It is therefore the value that gives rise to the average heating effect of the alternating current. The rms value of a current is literally the square root of the mean of the squares of all its varying current values over one complete cycle. Numerically, the rms value of a sinusoidal alternating current is its peak value divided by $\sqrt{2}$.

root-mean-square voltage (rms voltage) the numerical value given to an alternating voltage supply or potential difference that supplies the same power to a resistor as would be supplied by a ◊direct current of the same steady value. The rms value is literally the square root of the mean of the squares of all its varying voltages over one complete cycle. Numerically the rms value of a sinusoidal alternating voltage is the peak value divided by $\sqrt{2}$.

S

saturated solution solution obtained when a solvent (liquid) can dissolve no more of a solute (usually a solid) at a particular temperature. Normally, a slight fall in temperature causes some of the solute to crystallize out of solution. If this does not happen the phenomenon is called supercooling, and the solution is said to be ***supersaturated***.

scalar quantity in mathematics and science, a quantity that has magnitude but no direction, as distinct from a ◊vector quantity, which has a direction as well as a magnitude. Temperature, mass, and volume are scalar quantities.

scaler instrument that counts the number of radioactive particles passing through a radiation detector such as a Geiger–Muller tube (the scaler and tube together form a ◊Geiger counter). It gives the total, or cumulative, number of particles counted whereas a ◊ratemeter gives the number of particles detected in a unit of time.

scattering the random deviation or reflection of a stream of particles or of a beam of radiation such as light.

alpha particles scattered by a thin gold foil provided the first convincing evidence that atoms had very small, very dense, positive nuclei. From 1906 to 1908 Ernest Rutherford carried out a series of experiments from which he estimated that the closest approach of an alpha particle to a gold nucleus in a head-on collision was about 10^{-14} m. He concluded that the gold nucleus must be no larger than this. Most of the alpha particles fired at the gold foil passed straight through undeviated; however, a few were scattered in all directions and a very small fraction bounced back towards the source. This result so surprised Rutherford that he is reported to have commented: 'It was about as credible as if you had fired a 15-inch shell at a piece of tissue paper and it came back and hit you'.

light is scattered from a rough surface, such as that of a sheet of paper, by random reflection from the varying angles of each small part of the surface. This is responsible for the dull, flat appearance of such surfaces and their inability to form images (unlike mirrors). Light is also scattered by particles suspended in a gas or liquid. The red and yellow colours associated with sunrises and sunsets are due to the fact that red light is scattered to a lesser extent than is blue light by dust particles in the atmosphere. When the Sun is low in the sky, its light passes through a thicker, more dusty layer of the atmosphere, and the blue light radiated by it is scattered away, leaving the red sunlight to pass through to the eye of the observer.

seat belt safety device in a motor vehicle that is designed to reduce the risk of injury to a passenger during a collision or when brakes are applied sharply. In an emergency, it extends the time over which the decelerating force acts on a passenger thereby reducing that force to a safe level. It also spreads the force over a broad band across the chest and over the hip bone, reducing the pressure applied to the person.

The principle behind the operation of the seat belt is based on Newton's second law of motion (see ◊Newton's laws of motion). The change of momentum (◊impulse) required to stop a passenger is equal to the product of the decelerating force applied and the time over which that force acts. It follows that if the time is increased, the force will be reduced.

semiconductor crystalline material with an electrical conductivity between that of metals (good) and insulators (poor).

The conductivity of semiconductors can usually be improved by minute additions of different substances or by other factors. Silicon, for example, has poor conductivity at low temperatures, but this is improved by the application of light, heat, or voltage; hence silicon is used in transistors, rectifiers, and integrated circuits (silicon chips).

series circuit an electric circuit in which the components are connected end to end, so that the current flows through them all one after the other. The division of the ◊terminal voltage across each conductor is in the ratio of the resistances of each conductor. If the potential differences across two conductors of resistance R_1 and R_2, connected in

series circuit

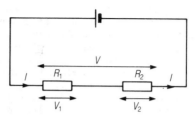

series, are V_1 and V_2 respectively, then the ratio of those potential differences is given by the equation:

$$V_1/V_2 = R_1/R_2$$

The total resistance R of those conductors is given by:

$$R = R_1 + R_2$$

Compare ◊parallel circuit.

shadow the area of darkness behind an opaque object that cannot be reached by some or all of the light coming from a light source in front. Its presence may be explained in terms of light rays travelling in straight lines and being unable to bend round obstacles. A point source of light produces an ◊umbra, a completely black shadow with sharp edges. An extended source of light produces both a central umbra and a ◊penumbra, a region of semidarkness with blurred edges where darkness gives way to light.

◊Eclipses are caused by the Earth passing into the Moon's shadow or the Moon passing into the Earth's shadow.

short circuit direct connection between two points in an electrical circuit. Its relatively low resistance means that a large current flows through it, bypassing the rest of the circuit, and this may cause the circuit to overheat dangerously.

short-sightedness or *myopia* defect of the eye in which a person can see clearly only those objects that are close up. It is caused by either the eyeball being too long or the cornea and lens system of the eye being too powerful, both of which cause the images of distant

shadow

point source of light

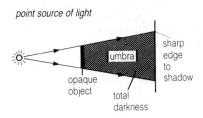

extended source of light

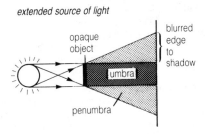

objects to be formed in front of the retina. Short-sightedness can be corrected by wearing spectacles fitted with ◊diverging lenses, or by wearing diverging (concave meniscus) contact lenses.

shunt conductor of very low resistance that is connected in parallel to an ◊ammeter in order to enable it to measure larger electric currents. Its low resistance enables it to act like a bypass, diverting the extra current through itself and away from the ammeter, and thereby reducing the instrument's sensitivity.

signal any sign, gesture, sound, or action that conveys information. Examples include the use of flags (semaphore), light (traffic and railway signals), radio telegraphy (Morse code), and electricity (telecommunications and computer networks).

silicon brittle non-metallic element (symbol Si) of proton number 14 and relative atomic mass 28.086. It is the second most abundant element (after oxygen) in the Earth's crust, but occurs only in combination with other elements—chiefly with oxygen to form silicates and silica, which makes up most sands and gravels.

The making of glass is based on the use of silica sands and dates back into prehistory. Today the crystalline form of silicon has become the basis of the electronics industry because of its ◊semiconductor properties, being used to make 'silicon chips' (◊integrated circuits).

silicon chip popular name for ◊integrated circuit.

siphon tube in the form of an inverted U with unequal arms. When it is filled with liquid and the shorter arm is placed in a tank or reservoir, liquid flows out of the longer arm provided that its exit is below the level of the surface of the liquid in the tank.

It works on the principle that the pressure at the liquid surface is atmospheric pressure, whereas at the lower end of the longer arm it is less than atmospheric pressure, causing flow to occur.

smoothing capacitor large ◊capacitor connected across the output of a rectifier circuit that has the effect of smoothing out the voltage variations to give a nearly steady DC voltage supply. The voltage and current output from a rectifier circuit fitted with a smoothing capacitor is similar to that provided by a battery.

Snell's law of refraction in optics, the law stating that for light rays passing from one transparent medium to another, the ratio of the sine of the ◊angle of incidence i and the sine of the ◊angle of refraction r is constant. The ratio for any pair of transparent media is called its ◊refractive index n. In equation terms:

$$n = \sin i / \sin r$$

solar energy energy derived from the Sun's radiation. The amount of energy falling on just 1 sq km is about 4,000 megawatts, enough to heat and light a small town. In one second the Sun gives off 13 million times more energy than all the electricity used in the USA in one year. *Solar heaters* have industrial or domestic uses. They usually consist of a black (heat-absorbing) panel containing pipes through which air or water, heated by the Sun, is circulated, either by thermal ◊convection currents or by a pump. Solar energy may also be harnessed indirectly using solar cells (photovoltaic cells) made of panels of ◊semiconductor material (usually silicon), which generate electricity when illuminated by sunlight. Although it is difficult to generate a high output from solar energy compared with sources such as nuclear or fossil-fuel energy, it is a major nonpolluting and renewable energy source used as far north as Scandinavia as well as in the SW USA and in Mediterranean countries.

A solar furnace, such as that built in 1970 at Odeillo in the French Pyrénées, has thousands of mirrors to focus the Sun's rays; it produces uncontaminated intensive heat for industrial and scientific or experimental purposes. Advanced schemes have been proposed that will use giant solar reflectors in space that would harness solar energy and beam it down to Earth in the form of ◊microwaves. Despite their low running costs, their high installation cost and low power output have meant that solar cells have found few applications outside space probes and artificial satellites. Solar heating is, however, widely used for domestic purposes in many parts of the world.

solar radiation radiation given off by the Sun, consisting mainly of visible light, ◊ultraviolet radiation, and ◊infrared radiation, although the whole spectrum of ◊electromagnetic waves is present, from radio waves to X-rays. High-energy charged particles such as electrons are also emitted, especially from solar flares. When these reach the Earth, they cause magnetic storms (disruptions of the Earth's magnetic field), which interfere with radio communications.

solar system the Sun and all the bodies orbiting it: the nine planets (Mercury, Venus, Earth, Mars, Jupiter, Saturn, Uranus, Neptune, and Pluto), their moons, the asteroids, and the comets. It is thought to have formed from a cloud of gas and dust in space about 4.6 billion years ago. The Sun contains 99% of the mass of the solar system. The edge of the solar system is not clearly defined, marked only by the limit of the Sun's gravitational influence, which extends about 1.5 light years, almost halfway to the nearest star, Alpha Centauri, 4.3 light years away.

solenoid elongated coil of wire. A strong and uniform magnetic field is produced inside it when a current passes through the wire. A solenoid fitted with an iron core forms an ◊electromagnet.

solid state of matter that holds its own shape (as opposed to a liquid, which takes up the shape of its container, or a gas, which totally fills its container). According to ◊kinetic theory, the atoms or molecules in a solid are not free to move but merely vibrate about fixed positions, such as those in crystal lattices.

solidification change of state from liquid to solid that occurs at the ◊freezing point of a substance.

solubility measure of the amount of solute (usually a solid or a gas) that will dissolve in a given amount of solvent (usually a liquid) at a particular temperature. Solubility may be expressed as grams of solute per 100 grams of solvent or, for a gas, in parts per million of solvent.

sonar acronym for *so*und *na*vigation and *r*anging A method of locating underwater objects by the reflection of ultrasonic waves. The time taken for an acoustic beam to travel to the object and back to the source enables the distance to be found since the velocity of sound in water is known. Sonar devices, or **echo sounders**, were developed 1920. Compare ◊radar.

sound physiological sensation received by the ear, originating in a vibration (pressure variation in the air) that communicates itself to the air, and travels in every direction, spreading out as an expanding sphere. All sound waves in air travel with a speed dependent on the temperature; under ordinary conditions, this is about 330 m per second. The pitch of the sound depends on the number of vibrations imposed on the air per second, but the speed is unaffected. The ◊loudness of a sound is dependent primarily on the amplitude of the vibration of the air.

The lowest note audible to a human being has a frequency of about 15–16 Hz (vibrations per second), and the highest one of about 20,000 Hz; the lower limit of this range varies little with the person's age, but the upper range falls steadily from adolescence onwards. Pressure waves of a frequency higher than the upper range are called ◊ultrasound.

sound absorption in acoustics, the conversion of sound energy to heat energy when sound waves strike a surface. The process reduces the amplitude of each reflected sound wave (echo) and thus the degree to which ◊reverberation takes place. Materials with good sound-absorbing properties are often fitted on walls and ceilings in buildings such as offices, factories, and concert halls in order to reduce or control sound levels.

sound-level indicator instrument used to measure the intensity or loudness of sound. Readings are given on a decibel (dB) scale that compares the sound level with the threshold of human hearing (standardized as an intensity of 1.0×10^{-12} watts per square metre). A specialized scale called the dBA scale gives a weighted reading that takes into account the ear's sensitivity to different frequencies.

sound wave the longitudinal wave motion with which sound energy travels through a medium. It carries energy away from the source of the sound without carrying the material itself with it. Sound waves are mechanical; unlike electromagnetic waves, they require vibration of their medium's molecules or particles, and this is why sound cannot travel through a vacuum.

source resistance alternative term for ◊internal resistance, the resistance inside an electric power supply.

specific heat capacity quantity of heat required to raise the temperature of unit mass (one kilogram) of a substance by one kelvin (1°C). The unit of specific heat capacity in the SI system is the joule per kilogram kelvin ($J\ kg^{-1}\ K^{-1}$).

specific latent heat the heat that changes the physical state of a unit mass (one kilogram) of a substance without causing any temperature change.

The *specific latent heat of fusion* of a solid substance is the heat required to change one kilogram of it from solid to liquid without any temperature change.

The *specific latent heat of vaporization* of a liquid substance is the heat required to change one kilogram of it from liquid to vapour without any temperature change.

spectrometer instrument used to study the composition of light emitted by a source. The range, or ◊spectrum, of wavelengths emitted by a source depends upon its constituent elements, and may be used to determine its chemical composition.

The simpler forms of spectrometer analyse only visible light. A *collimator* receives the incoming rays and produces a parallel beam, which is then split into a spectrum by either a diffraction grating or a prism mounted on a turntable. As the turntable is rotated each of the constituent colours of the beam may be seen through a *telescope*, and the angle at which each has been deviated may be measured on a circular scale. From this information the wavelengths of the colours of light can be calculated.

Spectrometers are used in astronomy to study the electromagnetic radiation emitted by stars and other celestial bodies. The spectral information gained may be used to determine their chemical composition, or to measure the ◊red shift of colours associated with the expansion of the universe and thereby calculate the speed with which distant stars are moving away from the Earth.

spectrum (plural *spectra*) arrangement of frequencies or wavelengths when electromagnetic radiations are separated into their constituent parts. Visible light is part of the ◊electromagnetic spectrum and most sources emit waves over a range of wavelengths that can be broken up or 'dispersed'; white light can be separated into red, orange, yellow, green, blue, indigo, and violet.

speed the rate at which an object moves. The constant speed v of an object may be calculated by dividing the distance s it has travelled by the time t taken to do so, and may be expressed as:

$$v = s/t$$

The usual units of speed are metres per second or kilometres per hour.

Speed is a scalar quantity in which direction of motion is unimportant (unlike the vector quantity ◊velocity, in which both magnitude and direction must be taken into consideration). See also ◊distance–time graph.

speed of light the speed at which light travels through empty space. Its value is 299,792,458 metres per second. The speed of light is the highest speed possible, and is independent of the motion of its source and of the observer. To accelerate any material body to this speed would require an infinite amount of energy.

speed of sound the speed at which sound travels through a medium, such as air or water. In air at a temperature of 0°C, the speed of sound is 331 metres per second. At higher temperatures, it is greater. It is greater in liquids and solids; for example, in water it is around 1,440 metres per second, depending on the temperature.

speed–time graph graph used to describe the motion of a body by illustrating how its speed or velocity changes with time. The gradient of the graph gives the object's acceleration: if the gradient is zero (the graph is horizontal) then the body is moving with constant speed or

speed-time graph

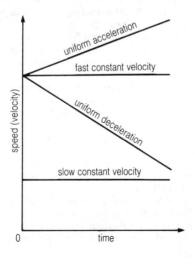

uniform velocity; if the gradient is constant, the body is moving with uniform acceleration. The area under the graph gives the total distance travelled by the body.

spring balance instrument for measuring weight that relates the weight of an object to the extent to which it stretches or compresses a vertical spring. According to ◊Hooke's law, the extension or compression will be directly proportional to the weight, providing that the spring is not overstretched. A pointer attached to the spring indicates the weight on a scale, which may be calibrated in newtons (the SI unit of force) for physics experiments, or in grams, kilograms, or pounds (units of mass) for everyday use.

stability measure of how difficult it is to move an object from a position of ◊equilibrium with respect to gravity.

An object displaced from equilibrium does not remain in its new position if its weight, acting vertically downwards through its ◊centre of mass, no longer passes through the line of action of the ◊contact force (the force exerted by the surface on which the object is resting), acting vertically upwards through the object's new base. If the lines of action of these two opposite but equal forces do not coincide they will form a couple and create a ◊moment that will cause the object either to return to its original rest position or to topple over into another position.

An object in *stable equilibrium* returns to its rest position after being displaced slightly. This form of equilibrium is found in objects that are difficult to topple over; these usually possess a relatively wide base and a low centre of mass—for example, a cone resting on its flat base on a horizontal surface. When such an object is tilted slightly its centre of mass is raised and the line of action of its weight no longer coincides with that of the contact force exerted by its new, smaller base area. The moment created will tend to lower the centre of mass and so the cone will fall back to its original position.

An object in *unstable equilibrium* does not remain at rest if displaced, but falls into a new position; it does not return to its original rest position. Objects possessing this form of equilibrium are easily toppled and usually have a relatively small base and a high centre of mass – for

stability

stable equilibrium

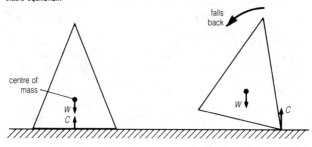

unstable equilibrium

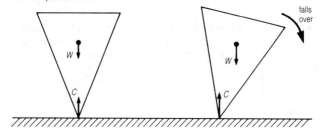

neutral equilibrium

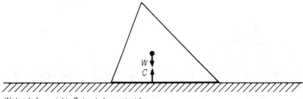

W stands for weight; *C* stands for contact force

example, a cone balancing on its point, or apex, on a horizontal surface. When an object such as this is given the slightest push its centre of mass is lowered and the displacement of the line of action of its weight creates a moment. The moment will tend to lower the centre of mass still further and so the object will fall on to another position.

An object in ***neutral equilibrium*** stays at rest if it is moved into a new position – neither moving back to its original position nor on any further. This form of equilibrium is found in objects that are able to roll, such as a cone resting on its curved side placed on a horizontal surface. When such an object is rolled its centre of mass remains in the same position, neither rising nor falling, and the line of action of its weight continues to coincide with the contact force; no moment is created and so its equilibrium is maintained.

stable equilibrium the state of equilibrium possessed by a body that will return to its original rest position if displaced slightly. See ◊stability.

standard form a method of writing numbers often used by scientists, particularly for very large or very small numbers. The numbers are written with one digit before the decimal point and multiplied by a power of 10. The number of digits given after the decimal point depends on the accuracy required. For example, the ◊speed of light is 2.9979×10^8 metres per second.

standing wave a wave in which the positions of ◊nodes (positions of zero vibration) and antinodes (positions of maximum vibration) do not move. Standing waves result when two similar waves travel in opposite directions through the same space.

For example, when a sound wave is reflected back along its own path, as when a stretched string is plucked, a standing wave is formed. In this case the antinode remains fixed at the centre and the nodes are at the two ends. Water and ◊electromagnetic waves can form standing waves in the same way.

state change change in the physical state (solid, liquid, or gas) of a material. For instance, melting, boiling, evaporation, and their opposites (solidification and condensation) are state changes.

state change

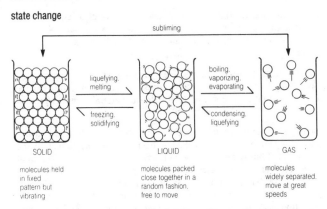

These changes require energy in the form of heat, called ◊latent heat, even though the temperature of the material does not change during the transition between states.

states of matter the forms (solid, liquid, or gas) in which material can exist. Whether a material is solid, liquid, or gas depends on its temperature and the pressure on it. The transition between states takes place at definite temperatures, called melting point and boiling point.

◊Kinetic theory describes how the state of a material depends on the movement and arrangement of its atoms or molecules. A hot ionized gas or plasma is often called the fourth state of matter, but liquid crystals, colloids, and glass also have a claim to this title.

static electricity ◊electric charge acquired by a body by means of electrostatic induction or friction. Rubbing different materials can produce static electricity, seen in the sparks produced on combing one's hair or removing a nylon shirt. In some processes static electricity is useful, as in paint spraying where the parts to be sprayed are charged with electricity of opposite polarity to that on the paint droplets.

statics branch of mechanics concerned with the behaviour of bodies at rest and forces in equilibrium, and distinguished from ◊dynamics.

stopping distance the minimum distance in which a vehicle can be brought to rest in an emergency from the moment that the driver notices danger ahead. It is the sum of the ◊thinking distance and the ◊braking distance.

stress and strain in the science of materials, measures of the deforming force applied to a body (stress) and of the resulting change in its shape (strain). For a perfectly elastic material, stress is proportional to strain (see ◊Hooke's law).

strings, vibrations of standing waves set up in stretched strings when they are plucked, or stroked with a bow. They are formed by the reflection of ◊progressive waves at the fixed ends of a string. Waves of many different ◊frequencies can be established on a string at the same time. Those that match the natural frequencies of the string will, by a process called ◊resonance, produce large-amplitude vibrations. The vibration of lowest frequency is called the ◊fundamental vibration; vibrations of frequencies that are multiples of the fundamental frequency are called harmonics.

sublimation change of state from solid to vapour, or from vapour to solid, that occurs without passing through an intermediate liquid state. In the formation of white frost, for example, water vapour in the atmosphere passes directly from the vapour state to the solid crystals of ice that make up frost.

surface tension the property that causes the surface of a liquid to behave as if it were covered with a weak elastic skin; this is why a needle can float on water. It is caused by the exposed surface's tendency to contract to the smallest possible area because of unequal cohesive forces between molecules at the surface. Allied phenomena include the formation of droplets, the concave profile of a meniscus, and the capillary action (see ◊capillarity) by which water soaks into a sponge.

suspension bridge bridge in which the spanning roadway or railway is supported by a system of steel ropes suspended from two or more tall towers.

switch device used to turn an electric current on and off, usually by closing or opening a circuit, or to change the route of a particular current. For example, switches are used to turn a light on or off, to select a channel on a television set, or to set the cooking time on an electric oven.

syringe cylindrical device with a nozzle at one end and a close-fitting piston at the other, used for injecting or extracting fluids (liquids or gases). It is filled by inserting the nozzle into the fluid, and then raising the piston; atmospheric pressure will then force the fluid into the low-pressure space created above the nozzle. Pushing the piston down will increase the pressure in the syringe, and force the fluid out once more.

T

tape chart speed–time graph constructed from a length of ticker-tape. See ♦ticker-timer.

tape recording, magnetic method of recording electric signals on a layer of iron oxide, or other magnetic material, coating a thin plastic tape. The electrical impulses are fed to the electromagnetic recording head, which magnetizes the tape in accordance with the frequency and amplitude of the original signal. The impulses may be audio (for sound recording), video (for television), or data (for computer). For playback, the tape is passed over the same, or another, head to convert magnetic into electrical signals, which are then amplified for reproduction. Tapes are easily demagnetized (erased) for reuse, and come in cassette, cartridge, or reel form.

telecommunications communications over a distance, generally by electronic means. Long-distance voice communication was pioneered in 1876 by Alexander Graham Bell, when he invented the telephone as a result of Michael Faraday's discovery of electromagnetism. Today it is possible to communicate with most countries by telephone cable, or by satellite or microwave link, with over 100,000 simultaneous conversations and several television channels being carried by the latest satellites. Integrated-services digital network (ISDN) is a system that transmits voice and image data on a single transmission line by changing them into digital signals, making videophones and high-quality fax possible; the world's first large-scale centre of ISDN began operating in Japan 1988. The chief method of relaying long-distance calls on land is microwave radio transmission.

telephone instrument for communicating by voice over long distances, invented by Alexander Graham Bell 1876. The transmitter (mouthpiece) consists of a carbon microphone, with a diaphragm that vibrates when a person speaks into it. The diaphragm vibrations com-

telecommunications

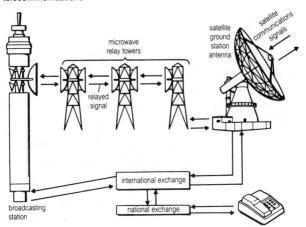

press grains of carbon to a greater or lesser extent, altering their resistance to an electric current passing through them. This sets up variable electrical signals, which travel along the telephone lines to the receiver of the person being called. There they cause the magnetism of an electromagnet to vary, making a diaphragm above the electromagnet vibrate and give out sound waves, which mirror those that entered the mouthpiece originally.

telescope device for collecting and focusing light and other forms of electromagnetic radiation from distant objects. A telescope produces a magnified image, which makes the object seem nearer, and it shows objects fainter than can be seen by the eye alone. A telescope with a large aperture, or opening, can distinguish finer detail and fainter objects than one with a small aperture. The *refracting telescope* uses lenses, and the *reflecting telescope* uses mirrors. A third type, the *catadioptric telescope*, with a combination of lenses and mirrors, is used increasingly. See also ◊radio telescope.

telephone

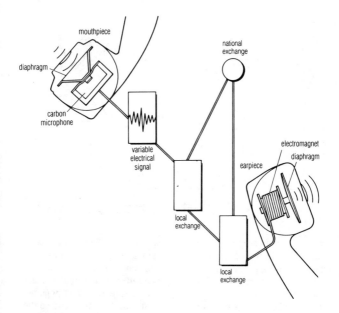

television (TV) reproduction at a distance by radio waves of visual images. For transmission, a television camera converts the pattern of light it takes in into a pattern of electrical charges. This is scanned line by line by a beam of electrons from an electron gun, resulting in variable electrical signals that represent the visual picture. These vision signals are combined with a radio carrier wave and broadcast. The TV aerial picks up the wave and feeds it to the receiver (TV set). This separates out the vision signals, which pass to a cathode-ray tube. The vision signals control the strength of a beam of electrons from an electron gun, aimed at the screen and making it glow more or less brightly.

telescope

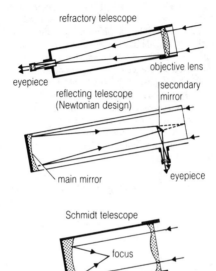

At the same time the beam is made to scan across the screen line by line, mirroring the action of the gun in the TV camera. The result is a re-creation of the pattern of light that entered the camera. In North America, 30 pictures are built up each second (25 in Europe), with a total of 525 lines in North America and Japan (625 lines in Europe).

temperature the state of hotness or coldness of a body, and the condition that determines whether or not it will transfer heat to, or receive heat from, another body according to the laws of thermodynamics. It is measured in degrees Celsius (before 1948 called centigrade), kelvin, or Fahrenheit.

The normal temperature of the human body is about 36.9°C.

temperature scales scale marked on a thermometer that gives a reading of temperature. The scales most widely used today are the ◊Celsius (or centigrade) scale, the ◊kelvin (or absolute) scale, and the Fahrenheit scale. Each scale has at least two fixed points that are used to calibrate all those thermometers using the scale. The fixed points most commonly used are the ice point and steam point of water. On the Celsius scale these are designated 0°C and 100°C respectively, the interval between them being divided equally into 100 units or degrees.

Other temperature scales include the platinum-resistance scale and the constant-volume gas-pressure scale. In physics, the term is also applied to the property of a material (for example, the expansion of mercury) chosen to indicate a change in temperature when constructing a particular type of thermometer

tension reaction force set up in a body that is subjected to stress. In a stretched string or wire it exerts a pull that is equal in magnitude but opposite in direction to the stress being applied at its ends. Tension originates in the net attractive intermolecular force created when a stress causes the mean distance separating a material's molecules to become greater than the equilibrium distance. It is measured in newtons.

terminal velocity or *terminal speed* the maximum velocity that can be reached by an object moving through a fluid (gas or liquid). As the speed of the object increases, so does the total magnitude of the forces resisting its motion. Terminal velocity is reached when the resistive forces exactly balance the applied force that has caused the object to accelerate; because there is now no resultant force, there can be no further acceleration. For example, an object falling through air will reach a terminal velocity and cease to accelerate under the influence of gravity when the air resistance equals the object's weight. Parachutes are designed to increase air resistance so that the acceleration of a falling person or package ceases more rapidly, thereby limiting terminal velocity to a safe level.

terminal voltage potential difference (pd) or voltage across the terminals of a power supply, such as a battery of cells. When the supply is not connected in circuit this voltage is the same as its ◊electromotive force (emf); however, as soon as it begins to supply current to a circuit its terminal voltage falls because some electric potential energy is lost in driving current against the supply's own ◊internal resistance. As the current flowing in the circuit is increased, the terminal voltage of the supply falls.

thermal capacity the heat energy, C, required to increase the temperature of an object by one degree. It is measured in joules per degree, J/°C or J/K. If an object has mass m and is made of a substance with ◊specific heat capacity c, then its thermal capacity C is given by:

$$C = mc$$

thermal expansion expansion that is due to a rise in temperature. It can be expressed in terms of linear, area, or volume expansion.

thermal reactor nuclear reactor in which the neutrons released by fission of uranium–235 nuclei are slowed down in order to increase their chances of being captured by other uranium–235 nuclei, and so induce further fission. The material (commonly graphite or heavy water) responsible for doing so is called a *moderator*. When the fast newly emitted neutrons collide with the nuclei of the moderator's atoms, some of their kinetic energy is lost and their speed is reduced. Those that have been slowed down to a speed that matches the thermal (heat) energy of the surrounding material are called *thermal neutrons*, and it is these that are most likely to induce fission and ensure the continuation of the chain reaction. See ◊nuclear reactor and ◊nuclear energy.

thermistor device whose electrical ◊resistance falls as temperature rises. The current passing through a thermistor increases rapidly as its temperature rises, and so it is used in electrical thermometers.

thermometer instrument for measuring temperature. There are many types, designed to measure different temperature ranges to varying degrees of accuracy. Each makes use of a different physical effect of temperature.

thermometer, clinical in medicine, a thermometer used to measure body temperature. It has a limited temperature range of 35–42°C, but

thermometer, clinical

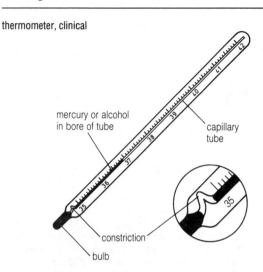

mercury or alcohol
in bore of tube

capillary
tube

constriction

bulb

gives readings in steps of 0.1 or 0.2 degrees. A narrow constriction in
the tube just above the bulb allows the thermometer to maintain a constant reading even after it has been removed from the patient: the mercury or alcohol is able to expand upwards into the bore of the tube as its
temperature rises, but cannot, as it cools, return to the bulb unless the
thermometer is shaken sharply.

thinking distance distance travelled by a vehicle from the moment
its driver notices danger ahead to the moment its brakes are applied.
Thinking distance is directly proportional to the speed of the vehicle,
and is also increased if the driver is tired or under the influence of alcohol. The total distance required to bring a vehicle to rest (its ◊stopping
distance) is the sum of the thinking distance and the ◊braking distance.

ticker-timer device used to time the motion of an object by printing
dots at regular time-intervals on a length of ticker-tape attached to that
object. When the timer is driven (through a transformer) by mains elec-

tricity, its printing head will vibrate with a frequency of 50 Hz, printing 50 dots a second on the tape. The movement of the object pulls the tape past the printing head, and as the speed of the object increases, the distance between adjacent dots grows wider. If the marked tape is cut up into lengths that represent the distance travelled by an object in a certain time (the object's speed), an experimental speed–time graph called a *tape chart* can be constructed, which can then be used to measure the object's acceleration and other aspects of its motion.

tidal energy energy derived from the tides. The tides gain their potential energy from the gravitational forces acting between the Earth and the Moon. If water is trapped at a high level during high tide, perhaps by means of a barrage across an estuary, it may then be gradually released and its associated ◊gravitational potential energy exploited to drive turbines and generate electricity. In the UK, several schemes have been proposed for the Bristol Channel, but environmental concerns as well as construction costs have so far prevented any decision from being taken.

tide rise and fall of sea level due to the gravitational forces of the Moon and Sun. High water occurs at an average interval of 12 hr 24 min 30 sec. The highest or *spring tides* are at or near new and full Moon; the lowest or *neap tides* when the Moon is in its first or third quarter. Some seas, such as the Mediterranean, have very small tides.

total internal reflection the complete reflection of a beam or ray of light at the interface of two transparent materials. No refraction takes place. Total internal reflection happens when: (1) the light ray is inside the optically more dense of the two materials, and (2) the ◊angle of incidence at the interface is greater than the ◊critical angle for that particular pair of materials.

Total internal reflection is used as a means of reflecting light inside ◊prisms and ◊optical fibres. Light is contained inside an optical fibre not by the cladding around it, but by the ability of the internal surface of the glass-fibre core to reflect 100% of the light, thereby keeping it trapped inside the fibre.

transformer device in which, by electromagnetic induction, an alternating current (AC) of one voltage is transformed to another voltage, without change of ◊frequency. Transformers are widely used in electrical apparatus of all kinds, and in particular in power transmission where high voltages and low currents are utilized.

A transformer has two coils, a primary for the input and a secondary for the output, wound on a common iron core. The ratio of the primary to the secondary voltages (and currents) is directly (and inversely) proportional to the number of turns in the primary and secondary coils.

If the numbers of turns in the primary and secondary coils are n_1 and n_2, the primary and secondary voltages are V_1 and V_2, and the primary and secondary currents are I_1 and I_2, then the relationships between these quantities may be expressed as:

$$V_1/V_2 = I_2/I_1 = n_1/n_2$$

transistor solid-state electronic component, made of ◊semiconductor material, with three or more ◊electrodes, that can regulate a current passing through it. A transistor can act as an amplifier, oscillator, photocell, or switch, and usually operates on a very small amount of power. Transistors commonly consist of a tiny sandwich of ◊silicon or germanium, alternate layers having different electrical properties. A crystal of pure silicon or germanium would act as an insulator (nonconductor).

By introducing impurities in the form of atoms of other materials (for example, boron, arsenic, or indium) in minute amounts, the layers may be made either **n-type** (*n*egative), having an excess of electrons, or **p-type** (*p*ositive), having a deficiency of electrons. This enables electrons to flow from one layer to another in one direction only.

Transistors have had a great impact on the electronics industry, and are now made in thousands of millions each year. They perform many of the same functions as the thermionic valve, but have the advantages of greater reliability, long life, compactness, and instantaneous action, no warming-up period being necessary. They are widely used in most electronic equipment, including portable radios and televisions, computers, satellites, and space research, and are the basis of the ◊integrated circuit (silicon chip). They were invented at Bell Telephone Laboratories in the USA in 1948 by John Bardeen and Walter Brittain, developing the work of William Shockley.

transmission of electrical power see ◊power transmission.

transverse wave

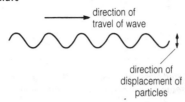

direction of
travel of wave

direction of
displacement of
particles

transverse wave wave in which the displacement of the medium's particles is at right angles to the direction of travel of the wave motion. It is characterized by its alternating crests and troughs. Simple water waves, such as the ripples produced when a stone is dropped into a pond, are transverse waves, as are the waves on a vibrating string.

All ◊electromagnetic waves have a transverse-wave form; their electric and magnetic fields (rather than the particles of their medium) vibrate at right angles to their direction of travel.

travelling wave another name for a ◊progressive wave.

truth table in electronics, a diagram representing the properties of a ◊logic gate.

TTL abbreviation for *transistor–transistor logic*, a family of integrated circuits with fast switching speeds commonly used in building electronic devices. TTL circuits require a very stable DC voltage of between 4.75 and 5.25 volts. See also ◊CMOS.

tuning fork simple metal instrument with two equal and parallel prongs, which vibrate with a single pure frequency or pitch when struck. It gives an accurate frequency that can be used to tune musical instruments, or as a standard reference in physics laboratories.

turbine engine in which steam, water, gas, or air is made to spin a rotating shaft by pushing on angled blades, like a fan. Turbines are among the most powerful of machines. Steam turbines are used to drive generators in power stations; water turbines spin the generators in hydroelectric power plants; and gas turbines (as jet engines) power most aircraft and drive machines in industry.

U

ultrasonics the study and application of the sound and vibrations produced by ultrasonic pressure waves (see ◊ultrasound).

ultrasound pressure waves similar in nature to sound waves but occurring at frequencies above 20,000 Hz (vibrations per second), the approximate upper limit of human hearing (15–16 Hz is the lower limit). Ultrasonics is concerned with the study and practical application of these phenomena.

ultraviolet radiation light rays invisible to the human eye, of wavelengths from about 4×10^{-7} to 5×10^{-9} metres (where the ◊X-ray range begins). Physiologically, they are extremely powerful, producing sunburn and causing the formation of vitamin D in the skin.

Ultraviolet rays are strongly germicidal and may be produced artificially by mercury vapour and arc lamps for therapeutic use. The radiation may be detected with ordinary photographic plates or films down to 2×10^{-6} metres. It can also be studied by its fluorescent effect on certain materials. The desert iguana, *Disposaurus dorsalis*, uses it to locate the boundaries of its territory and to find food.

umbra region of a ◊shadow that is totally dark because no light reaches it, and from which no part of the light source can be seen (compare ◊penumbra). In astronomy, it is a region of the Earth from which a complete ◊eclipse of the Sun or Moon can be seen.

unit standard quantity in relation to which other quantities are measured. There have been many systems of units. Some ancient units, such as the day, the foot, and the pound, are still in use. SI units, the latest version of the metric system, are widely used in science.

unstable equilibrium the state of equilibrium possessed by a body that will not remain at rest if displaced slightly, but will topple over into a new position; it will not return to its original rest position. See ◊stability.

upthrust upwards force experienced by all objects that are totally or partially immersed in a fluid (liquid or gas). It acts against the weight of the object, and, according to Archimedes' principle, is always equal to the weight of the fluid displaced by that object. An object will float when the upthrust from the fluid is equal to its weight. See ◊floating.

uranium radioactive metallic element (symbol U) of proton number 92 and relative atomic mass 238.029. It is the most abundant radioactive element in the Earth's crust, its decay giving rise to essentially all the radioactive elements in nature; its final decay product is the stable element lead. Uranium is one of the three elements capable of ◊nuclear fission (the other two are plutonium and thorium); the isotope uranium-235 is used as a fuel in nuclear reactors and in making nuclear weapons.

U-tube U-shaped tube that may be partly filled with liquid and used as a manometer, or instrument for measuring ◊pressure. Greater pressure on the liquid surface in one arm will force the level of the liquid in the other arm to rise. A difference between the pressures in the spaces in

U-tube

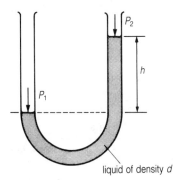

liquid of density *d*

the two arms is therefore registered as a difference in the heights of the liquid in the arms. If the greater pressure is P_1 pascals and the lesser pressure is P_2 pascals, then:

$$P_1 = P_2 + hdg$$

where h is the difference (in metres) in the heights of the liquid, d is the density (in kilograms per cubic metre) of that liquid, and g is the gravitational field strength (in newtons per kilogram). If P_2 is atmospheric pressure, then hdg gives the pressure difference between P_1 and the atmosphere.

U-value measure of a material's heat-conducting properties. It is used in the building industry to compare the efficiency of insulating products, a good insulator having a low U-value. The U-value of a material is defined as the rate at which heat is conducted through it per unit surface area per unit temperature-difference between its two sides; it is measured in watts per square metre per kelvin (W m^{-2} K^{-1}). In equation terms:

$$\text{U-value} = \frac{\text{rate of loss of heat}}{(\text{surface area} \times \text{temperature difference})}$$

V

vacuum in general, a region completely empty of matter; in physics, any enclosure in which the gas pressure is considerably less than atmospheric pressure (101,325 pascals).

van de Graaff generator electrostatic generator capable of producing a voltage of over a million volts. It consists of a continuous vertical conveyor belt that carries electrostatic charges (resulting from friction) up to a large hollow sphere supported on an insulated stand. The lower end of the belt is earthed, so that charge accumulates on the sphere. The size of the voltage built up in air depends on the radius of the sphere, but can be increased by enclosing the generator in an inert atmosphere, such as nitrogen.

vaporization change of state of a substance from liquid to vapour. See ◊evaporation.

vapour one of the three states of matter (the other two are ◊solid and ◊liquid). The molecules in a vapour move randomly and are far apart, the distance between them, and therefore the volume of the vapour, being limited only by the walls of any vessel in which they might be contained. A vapour differs from a ◊gas only in that a vapour can be liquefied by increased pressure, whereas a gas cannot unless its temperature is lowered below its ◊critical temperature; it then becomes a vapour and may be liquefied.

vector quantity any physical quantity that has both magnitude and direction (such as the velocity or acceleration of an object) as distinct from a scalar quantity, which has magnitude but no direction (such as speed, density, or mass). A vector may be represented geometrically by an arrow whose length corresponds to its magnitude, pointing in an appropriate direction. Vectors can be added graphically by constructing a parallelogram of vectors (such as the ◊parallelogram of forces).

van de Graff generator

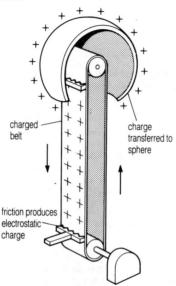

charged belt

charge transferred to sphere

friction produces electrostatic charge

velocity the speed of an object in a given direction, or the rate of change of an object's displacement. The velocity v of an object travelling in a fixed direction may be calculated by dividing the distance s it has travelled by the time t taken to do so, and may be expressed as:

$$v = s/t$$

The usual units of speed are metres per second or kilometres per hour. See also ◊distance–time graph and ◊equations of motion.

Velocity is a ◊vector quantity, since its direction is as important as its magnitude (or speed). If the direction of motion of a body changes, even if it is travelling at constant speed, then its velocity is also changing and it is therefore accelerating. The velocity at any instant of a particle travelling in a curved path is in the direction of the tangent to the path at the instant considered.

velocity ratio (VR) or *distance ratio* in a machine, the distance moved by the input force, or effort, divided by the distance moved by the output force, or load in the same time. It follows that the velocities of the effort and the load are in the same ratio. Velocity ratio has no units. See also ◊efficiency.

velocity, uniform motion of an object travelling with constant speed and in a constant direction. An object that has uniform velocity covers equal distances in the same direction in equal successive time intervals. See ◊speed–time graph.

VHF (abbreviation for *very high frequency*) referring to radio waves that have very short wavelengths. They are used for interference-free FM transmissions (see ◊frequency modulation). VHF transmitters have a relatively short range because the waves cannot be reflected over the horizon like longer radio waves.

volt SI unit (symbol V) of electromotive force or electric potential. A small battery has a potential of 1.5 volts; the domestic electricity supply in the UK is 240 volts (110 volts in the USA); and a high-tension transmission line may carry up to 500 kilovolts.

voltage term commonly used for ◊potential difference (pd).

voltage amplifier electronic device that increases an input signal in the form of a voltage or ◊potential difference, delivering an output signal that is larger than the input by a specified ratio.

voltage comparator electronic circuit that compares two input voltages or ◊potential differences, giving an output voltage that is proportional to the difference between the two inputs.

voltage divider alternative term for ◊potential divider.

voltmeter instrument for measuring potential difference (voltage). It has a high internal resistance (so that it passes only a small current), and is connected in parallel with the component across which potential difference is to be measured. A common type is constructed from a sensitive current-detecting ◊moving-coil meter placed in series with a high-value resistor (⬦multiplier). To measure an AC (alternating current) voltage, the circuit must usually include a ⬦rectifier; however, a moving-iron instrument can be used to measure alternating voltages without the need for such a device.

VR abbreviation for ⬦*velocity ratio*.

W

watt SI unit (symbol W) of power (the rate of expenditure or consumption of energy). A light bulb may use 60, 100, or 150 watts of power; an electric heater will use several kilowatts (thousands of watts).

The watt is defined as the power used when one joule of work is done in one second. In electrical terms, the flow of one ampere of current through a conductor whose ends are at a potential difference of one volt uses one watt of power (watts = volts × amps). The unit is named after the Scottish engineer James Watt.

wave disturbance travelling through a medium (or space). There are two types: in a *longitudinal wave* (such as a ◊sound wave) the disturbance is parallel to the wave's direction of travel; in a *transverse wave* (such as an ◊electromagnetic wave) it is perpendicular. The medium (such as the Earth, in the case of seismic waves) is not permanently displaced by the passage of a wave.

wave energy energy derived from that of water waves. Various schemes have been advanced since 1973, when oil prices rose dramatically and an energy shortage threatened. In 1974 the British engineer Stephen Salter developed the duck – a floating boom whose segments nod up and down with the waves. The nodding motion can be used to drive pumps and spin generators. Another device, developed in Japan, uses an oscillating water column to harness wave energy.

wave equation equation relating the speed of a wave to its frequency and wavelength. If the wave's speed is c, its frequency f, and its wavelength λ, then:

$$c = f\lambda$$

wave front imaginary line joining a set of particles that are in ◊phase (in step) in a wave motion. It is, in effect, the shape of a wave as seen from above the plane in which it moves. All the particles along the crest of a wave are in phase and can therefore be considered to form a wave front.

wavelength the distance between successive crests of a ◊wave. The wavelength of a light wave determines its colour; red light has a wavelength of about 700 nanometres, for example. The complete range of wavelengths of electromagnetic waves is called the electromagnetic ◊spectrum.

weight the force exerted on an object by ◊gravity. The weight of an object depends on its mass – the amount of material in it – and the Earth's gravitational field strength. If the mass of a body is m kilograms and the gravitational field strength is g newtons per kilogram, its weight W in newtons is given by:

$$W = mg$$

The strength of the Earth's gravitational field strength decreases with height; consequently, an object will weigh less at the top of a mountain than at sea level. On the Moon, an object weighs only one-sixth of its weight on Earth because the pull of the Moon's gravity is one-sixth that of the Earth.

weightlessness condition in which there is no gravitational force acting on a body, either because gravitational force is cancelled out by equal and opposite acceleration, or because the body is so far outside a planet's gravitational field that no force is exerted upon it.

wheel and axle simple machine with a rope wound round an axle connected to a larger wheel with another rope attached to its rim. Pulling on the wheel rope (applying an effort) lifts a load attached to the axle rope. The velocity ratio of the machine (distance moved by load divided by distance moved by effort) is equal to the ratio of the wheel radius to the axle radius.

wind energy energy derived from the wind. It is harnessed by sailing ships and windmills, both of which are ancient inventions, and by wind ◊turbines, aerodynamically advanced windmills that drive electricity generators when their blades are spun by the wind. Wind energy is a renewable resource that produces no direct pollution of the air; it is therefore beginning to be used to produce electricity on a large scale.

work measure of the result of transferring energy from one system to another to cause an object to move. Work should not be confused with ◊energy (the capacity to do work, which, like work, is measured in joules) or with ◊power (the rate of doing work, measured in joules per second).

Work is equal to the product of the force used and the distance moved by the object in the direction of that force. If the force is F newtons and the distance moved is d metres, then the work W is given by:

$$W = Fd$$

For example, the work done when a force of 10 newtons moves an object 5 metres against some sort of resistance is 50 joules (50 newton metres).

X

X-ray band of electromagnetic radiation in the wavelength range 10^{-11} to 10^{-9} m (between gamma rays and ultraviolet radiation; see ◊electromagnetic waves). Applications of X-rays make use of their short wavelength (as in X-ray crystallography) or their penetrating power (as in medical X-rays of internal body tissues). X-rays are dangerous and can cause cancer.

X-rays were discovered by Wilhelm Röntgen in 1895 and formerly called roentgen rays. They are produced when high-energy electrons from a heated filament cathode strike the surface of a target (usually made of tungsten) on the face of a massive heat-conducting anode, between which a high alternating voltage (about 100 kV) is applied.

X-ray

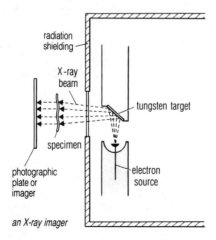

radiation
shielding

X-ray
beam

tungsten target

specimen

electron
source

photographic
plate or
imager

an X-ray imager

Appendix I

acceleration

$$a = \frac{v - u}{t}$$

$$a = \frac{F}{m} \quad \text{(Newton's second law of motion)}$$

Boyle's law

$$P_1 V_1 = P_2 V_2$$

centripetal force

$$F = \frac{mv^2}{r}$$

charge

$$Q = \frac{W}{V}$$

$$Q = It$$

Charles's law

$$\frac{V_1}{T_1} = \frac{V_2}{T_2}$$

current

$$I = \frac{Q}{t}$$

$$I = \frac{V}{R} \quad \text{(Ohm's law)}$$

$$\frac{I_1}{I_1} = \frac{R_2}{R_2} \quad \text{(parallel circuit)}$$

density

$$D = \frac{m}{V}$$

efficiency

$$\text{efficiency} = \frac{\text{useful work output}}{\text{work input}} \times 100\%$$

$$\text{efficiency} = \frac{\text{useful power output}}{\text{power input}} \times 100\%$$

$$\text{efficiency} = \frac{\text{mechanical advantage}}{\text{velocity ratio}} \times 100\%$$

electrical energy

$$W = QV$$
$$W = I^2Rt$$

electromotive force

$$E = V + Ir$$
$$E = I(R + r)$$

equations of motion

$$v = u + at$$
$$s = \frac{1}{2}(u + v)t$$
$$s = vt + \frac{1}{2}at^2$$
$$s = ut + \frac{1}{2}at$$
$$v^2 = u + 2as$$

force

$$F = ma \quad \text{(Newton's second law of motion)}$$

gain

$$\text{gain} = \frac{\text{amplitude of output signal}}{\text{amplitude of input signal}} \quad \text{(amplifier)}$$

$$\text{voltage gain} = \frac{\text{output voltage}}{\text{input voltage}} \quad \text{(voltage amplifier)}$$

$$\text{voltage gain} = \frac{\text{resistance of feedback resistor}}{\text{resistance of input resistor}} \quad \begin{array}{l}\text{(operational}\\\text{amplifier)}\end{array}$$

gas law, general

$$\frac{P_1 V_1}{T_1} = \frac{P_2 V_2}{T_2}$$

gravitational field strength

$$F = G\frac{m_1 m_2}{r^2} \quad \text{(Newton's universal law of gravitation)}$$

gravitational potential energy

$$E_g = mgh$$

impulse

$$J = mv - mu$$
$$J = Ft$$

kinetic energy

$$E = \tfrac{1}{2}mv^2$$

magnification

$$\text{magnification} = \frac{h_\text{I}}{h_\text{O}}$$

$$\text{magnification} = \frac{d_\text{I}}{d_\text{O}}$$

mass

$$m = \frac{F}{a}$$

$$m = \frac{W}{g}$$

mechanical advantage

$$\text{mechanical advantage} = \frac{\text{load}}{\text{effort}}$$

moment

$$\text{moment} = Fd$$

momentum

$$\text{momentum} = mv$$

potential difference

$$V = \frac{W}{Q}$$

$$V = IR \qquad \text{(Ohm's law)}$$

potential divider

$$V_\text{out} = V_\text{in} \frac{R}{R + R}$$

power

$$P = \frac{W}{t}$$

power, electrical

$$P = IV$$
$$P = I^2R$$

pressure

$$P = \frac{F}{A}$$
$$P = hdg \qquad \text{(in a liquid or gas)}$$
$$P_1 = P_2 + hdg \qquad \text{(in a U-tube)}$$

pressure law

$$\frac{P_1}{T_1} = \frac{P_2}{T_2}$$

refractive index

$$n = \frac{\sin i}{\sin r}$$

resistance

$$R = \frac{V}{I} \qquad \text{(Ohm's law)}$$
$$R = R_1 + R_2 + \ldots \qquad \text{(resistors in series)}$$
$$\frac{1}{R} = \frac{1}{R_1} + \frac{1}{R_2} + \ldots \qquad \text{(resistors in parallel)}$$

resistivity

$$\rho = \frac{RA}{l}$$

speed

$$v = \frac{s}{t}$$

thermal capacity

$$C = mc$$

transformer

$$\frac{V_1}{V_2} = \frac{I_2}{I_1} = \frac{n_1}{n_1}$$

U-value

$$\text{U-value} = \frac{\text{rate of loss of heat}}{\text{surface area} \times \text{temperature difference}}$$

velocity

$$v = \frac{s}{t}$$

velocity ratio (distance ratio)

$$\text{velocity ratio} = \frac{\text{distance moved by effort}}{\text{distance moved by load}}$$

wave speed

$$c = f\lambda \qquad \text{(wave equation)}$$

weight

$$W = mg$$

work

$$W = Fd$$

Appendix II
SI units and multiples

quantity	SI unit	symbol
absorbed radiation dose	gray	Gy
amount of substance	mole*	mol
electric capacitance	farad	F
electric charge	coulomb	C
electric conductance	siemens	S
electric current	ampere*	A
energy or work	joule	J
force	newton	N
frequency	hertz	Hz
illuminance	lux	lx
inductance	henry	H
length	metre*	m
luminous flux	lumen	lm
luminous intensity	candela*	cd
magnetic flux	weber	Wb
magnetic flux density	tesla	T
mass	kilogram*	kg
plane angle	radian	rad
potential difference	volt	V
power	watt	W
pressure	pascal	Pa
radiation dose equivalent	sievert	Sv
radiation exposure	roentgen	r
radioactivity	becquerel	Bq
resistance	ohm	W
solid angle	steradian	sr
sound intensity	decibel	dB
temperature	°Celsius	°C
temperature, thermodynamic	kelvin*	K
time	second*	s

*SI base unit

SI prefixes

multiple	prefix	symbol	example
1,000,000,000,000,000,000 (10^{18})	exa-	E	Eg(exagram)
1,000,000,000,000,000 (10^{15})	peta-	P	PJ (petajoule)
1,000,000,000,000 (10^{12})	tera-	T	TV (teravolt)
1,000,000,000 (10^9)	giga-	G	GW (gigawatt)
1,000,000 (10^6)	mega-	M	MHz (megahertz)
1,000 (10^3)	kilo-	k	kg (kilogram)
100 (10^2)	hecto-	h	hm (hectometre)
10	deca-	da-	daN (decanewton)
1/10 (10^{-1})	deci-	d	dC(decicoulomb)
1/100 (10^{-2})	centi-	c	cm(centimetre)
1/1,000 (10^{-3})	milli-	m	mA(milliampere)
1/1,000,000 (10^{-6})	micro-	µ	µF (microfarad)
1/1,000,000,000 (10^{-9})	nano-	n	nm(nanometre)
1/1,000,000,000,000 (10^{-12})	pico-	p	ps (picosecond)
1/1,000,000,000,000,000 (10^{-15})	femto-	f	frad(femtoradian)
1/1,000,000,000,000,000,000 (10^{-18})	atto-	a	aT(attotesla)